# Conservation of the *Homo sapiens*:

## On the Cybernetics of education;

An evolutionary third-order Cybernetic perspective of the
education for sustainable development

Gihan S. Soliman, PGCE

gihansami@yahoo.com

284678811©1/3/2014 UK Copyright Registration Service

International-Curricula Educators Association

ICEA

www.icea-global.org

# TABLE OF CONTENTS

"Survival of our species is linked to education" (Curzon, 1990).
"The whole is something over and above its parts and not just the sum of them all..." Aristotle

## 1- EDUCATION AND. LEARNING

Education and cognition are not synonymous.

Education is a "distributed cognition" (Jones & Nemeth, 2005) that is of a potential benefit for both the individual and the communities. Subsequently, it cannot occur naturally or neutrally for two simple reasons; firstly, in its most liberated form, education would be a supervised exploration of cognitive artefacts (Jones and Nemeth, 2005). And secondly, because "distributed cognition "usually gets assessed for efficiency and effectiveness, so as to satisfy the needs of the community chiefly for the purpose of certification. For this and more; education is multi-stakeholder (Reece and Walker, 2007).

This discourse is essential to segregate between learning theories and teaching theories (Holmes and Abington-Cooper, 2000). In planning learning and assessment, a teacher may benefit from the application and integration of the different psychological and neurological theories on learning (Reece and Walker, 2007) yet, draws on other important concepts, theories and skills such as communication models, curriculum models, assessment principles, administrative skills, system cybernetics and management cybernetics.

## 2- WHY EDUCATE?

Darwin (1871) who considered human beings as a species of mammals, was bewildered how, unlike all mobile organisms on earth, we learn to survive, coexist and cooperate mainly through culture rather than the accumulation of genetic trait alteration across generations (Darwin, 1871). Some animals might, in relatively rare cases, learn some new tricks by imitation (i.e. culture); but such learning would not be necessary for their survival at the time of learning. Whatever is critical for their survival and coexistence had been meticulously encoded into their genes through the long slow double-loop process of evolution, or will be. A salmon who has never recognized parents will set off for a long trip to the sea then comes back to her "birthplace" in the river to lay her eggs, then inevitably dies before meeting the offspring or get the chance to nurture them (Ridly,1995). A male Emperor penguin stands on the beach for nearly two months holding an egg on its feet till it hatches then sets off for feeding (Encyclopaedia Britannica), and a predator cannot avoid killing for survival. Retrospectively, animals have no moral standards; no "right" or "wrong", they are rather "just what they are" (Darwin, 1871; Glasersfeld, 2002).

While in human societies, education is ideally expected to "promote the emotional, intellectual, social and economic wellbeing of individuals and populations as a whole" (Domain A, LLUK professional Standards).That is not to suggest that *Homo sapiens* are the only species cable of learning through culture but that *Homo sapiens* are the only species who *have to* learn mainly through culture.

At a certain point of our evolution, social standards (traits underlying tendencies of various social behaviour) required for our coexistence and cooperation seemed to have mostly been removed from the realm of genetics that we had to reinvent them for and through culture (Darwin, 1871; Childe, 1936, cited in Corning, 1997). This "cultural transmission" required for our "biological make up"(Gray, 2011) is called education. Moreover, at some point of community development; some 10,000 years ago, a need for mass education originated due to the invention of agriculture and the use of artefacts.

The cognitive artefacts (Jones and Nemeth, 2005) resulted in the accumulation of a huge heritage of knowledge, sciences, ethics and arts, that the "hunter-gatherer style" of learning (Gray, 2011) was no longer functional to allow community development. Sharing sophisticated knowledge had become, from that era on, a requirement of survival, for both the individuals and groups (Jones & Nemeth, 2005).

A great deal of our evolutionary traits, since then, are currently encoded into texts and cybernetic models rather than our genome, and with the accumulation of such cognitive codes, we have been able to fly in the sky and dive into the sea, crossing vast deserts and oceans; dressed up like wild flowers in summer and polar bears in winter.

Paradoxically, the attempts to understand why and how people learn, were usually initiated by biologists, neurologists or psychotherapists but rarely, if ever, by educators. Moreover, such theories were developed through "establishing individual territories and specialisms", where each theorist tried to stay distinct from his/her progenitors and critical to their works (Coffield et al, 2004) rather than attempting to integrate knowledge towards a holistic approach. Subsequently, theories are usually studied by teachers-to-be as independent of each other; each with a distinctive set of pioneers, critiques, advantages and limitations (Reece & Walker, 2007).

Reece and Walker (2007) provide a very useful insight of the five main movements dominant in education, articulated as: the Behaviourism, the Neo-behaviourism, the Gestalt, the Cognitivism/constructivism and the Humanistic education.

To sum up the theories; Behaviourism is a psychotherapeutic theory based on a stimulus-response (S_R) concept, conducted experimentally on animals (and sometimes a toddler), proposing the possibility of predicting the human behaviour by imposing a certain sequence of stimuli and reinforcement (conditioning). Under Behaviourism comes also the social learning paradigm (Bandura, 1971). While Neo-behaviourism is a less extreme version of behaviourism based on the mere preference for using evidence-based approaches in dealing with human behaviour rather than philosophical ones. Cognitivism/constructivism, on the other hand, does not concern itself with the behaviour of the students as much as it is focused on how the learners construct and organize their knowledge reservoir, while Humanistic education is mainly focused on allowing/facilitating self-motivated, self-directed learning towards "self- fulfilment" (in Curzon, L.B. 1990; Reece & Walker, 2007). Under the humanistic approach, there are another two important concepts significant for our study, firstly, the "Gestalt" theory which proposes that the whole is "greater than the sum of its parts"(Aristotle; Soff, M. 2013) and secondly, the second order cybernetics as a cognitive theory (Glasersfeld, 2002).

Between educational theories provided mostly by non-educators and therapeutic theories applied by non-therapists (Peel, 2005), Cybernetics as the art of "steermanship"(Weiner,1948 cited in Mindell, 2000) provides a key for understanding and managing the complexity of the educational process, rarely acknowledged.

## 3- CYBERNETICS AND LIFE

Cybernetics is from the Greek word *kybernan* meaning to steer or to govern. The term has been introduced in 1948 by Norbert Weiner to describe how systems work in machines and animals and is commonly known as "the scientific study of how people, animals, and machines control and communicate information" (Merriam-Webster, 2014) and therefore is relevant to all sorts of life interactions (Cornell, 1997, Maturana and Valera, 1928; Galsersfeld, 2000, Umpleby, 2007).

Cybernetics has been regarded as a difficult concept to grasp because permeates all knowledge domains and is frequently expressed in different jargons, which are all correct in principle yet un-matching and confusing in case of cross speciality reading or research. However Cybernetics is simply about the unique intellectual ability of human beings to observe, describe and/or simulate self-regulatory systems constituting life (Corning, 1997; ASC) on earth.

A recent study by Franz-Xaver Neubert - Oxford University (Nature World News, 2014) has identified an area in the ventrolateral frontal cortex region of the brain, called the Lateral frontal pole prefrontal cortex, that seems to be "uniquely human" and "does not seem to have an equivalent in the brains of monkeys at all which is involved in strategic planning, decision making and multitasking abilities"(BBC News). Strategic planning certainly involves cybernetics. This scientific discovery had been unconsciously foreshadowed in Darwin's (1871) The *Descent of Man* by a sarcastic remark related to the idea of classifying the *Homo sapiens* as a separate biological kingdom, which he believed would have never been proposed, if *Homo sapiens* were not themselves the taxonomists! An observation suggestive of a unique human intellectual ability of observation: Who else is capable of, or takes an interest in observing life systems, give names or classify species and why?

In his Holistic Darwinism, Corning(1997) sheds the light on the true nature of cybernetics in relation to life systems in a teleological approach, wrapping up a huge compilation of definitions provided by hundreds of scientists, philosophers, institution and Cyberneticists in several domains of knowledge and areas of specialism, such as Weiner; Stafford Beer, Glasersfeld, Umpleby (1982), Richard (1999), ASC constitution, ASC (1987), Ampere, Ashby, Gregory Batson, Maturana and Valera (1928) and others:

"The science of cybernetics is not about thermostats or machines; that characterization is a caricature. Cybernetics is about purposiveness, goals, information flows, decision-making control processes and feedback (properly defined) at all levels of living systems." (Corning, 1997).

## 4- WHICH IS THE SYSTEM?

Scientists have proposed that "the universe, matter, molecules, cells, and ecosystems, among other aspects of nature, are self-organizing". (Maturana and Valera, 1928; Umpleby, 2007). Expressed differently, based on the relation between information, energy and matter, that a life system is a bundle of information, energy and/or matter [IEM] (ibid).

I, on the other hand, will use the acronym **hICEM** the communication within the system and in/outward - for the purpose of information flow analysis - as an interdependent aspect of the system.

**ICEM = I + [Cx+Cn] + [E] ± M**

**[Cx]:** Extrinsic communication trajectory.
**[Cn]:** Intrinsic communication trajectory.

**[h]** : Relativity factor – the wholeness of the system as perceived by an observer at a certain point of time and space.

This is not a mathematical formula but rather an illustrative one. It is significant in approaching real-life system or ecosystems as it expresses the deterministic structural (Matuarana and Valera, 2012) relationship of a focal system with (an)other system(s) - without losing focus on the focal system -   and spatiotemporal factors in relationship to an observer.

Cybernetic systems need communication for control to be exercised. There is no feedback that is not communicative in intent, and a control intent (to permit the shorthand of first-order Cybernetics) has also to be communicated. Communication is, therefore, necessary to the exercise of control, and therefore to cybernetic systems.

An illustration of the significance of examining the role of communication as an interdependent component of any given system rather than a virtual linkage between its components is in the authentic existence of the RNA within the composition of a gene (Ophardet, 2003) and the existence of the endocrine system, the nervous system as well as the sensory system in human beings and animals. Communication is beyond the matter/energy of the system and is not always manifested in either of them (Glaneville, 2004). More precisely, control is rather in the wholeness of the system or the alignment of such natural elements in such a manner which allows such intrinsic and extrinsic communication to occur under such natural forces at such a time and such a position in the space as perceived by an observer. Such wholeness producing such dynamics of self-governance is always "greater than the sum of its fragments" (Aristotle cited in Corning, 1997). Therefore, studying a system without depicting communication as an interdependent aspect must lead to erratic coding of the model as it ignores the system's fundamental relation to the environment and its impact on the systems and overlooks the system's learning capacity.

Energy and matter are inter-reversible and often treated as "one entity" in biology, while in physics the law regulating them is E=mc2. Information however is the "difference which makes the difference" and - unlike energy and matter - is known to be the "role of an observer" (ibid). This is where the huge difference between a biological perspective of systems and a physical one comes. Biology while trying to commits itself to objectivity ignores the role of the observer which defines the precise relationship between matter and energy at a certain point of time and space as perceivable by an observer and thus ignores the system-creation piece of information which is necessarily beyond its testable substance. This means that biology will always produce an incomplete image of the reality as perceived by an observer. It seems that biological sciences tend to consider information as the matter/energy while physics admits it is above them both (as obvious from the equation). For decades, biologists have denied or resisted the systematic interpretation of Darwin and his holistic approach to species (as intergenerational populations) and kept looking for system control in their fragments till a holistic Darwinism has been acknowledged. However, the heritage of fragmentation is heavy and there is so much work to do to cut across specialities and address the gap between sciences through second-order cybernetics.

The difference between a living system and a self-regulatory machine would basically lie in the amazing ability of the living system generate its own  energy/matter required for self-generated motion through subsuming other organic being, while the machine, in spite of its "built in" autonomy, can only operate through provided energy/matter, and provided purpose. The purpose of an observer (Corning, 1947; Galsersfeld, 2000).

In studying the gene, we can make out its components as the defining information represented by the DNA and the communication represented by the RNA and the protein which represents the energy/matter It is relatively easy to biologists to study how life systems complexify through the function of this amazing cybernetic unit and come up with innovative cloning or mutation tricks, but the answer to what originates such alignment of such natural elements so as to allow such performance is a piece of information usually overlooked by biologists as it perhaps involves too much uncertainty for science to handle. What makes this tiny whole - that is greater than the sum of its parts - a functional whole?

The stability of this system (thermostat) does not exist either in the sensor/switch or in the heater. It lies between them. It is the whole system that is stable achieving the desired constant temperature. (Ranulph Glanville).

## AN ILLUSTRATION OF A REAL LIFE-SYSTEM

$[E = mc^2]$ ==>

HhIECM = Relativity factor [Hh] X (I + [Cx+Cn] + [E] ± M)

HhICEM=Hh X [I.a+I.a~b+I.a~b~c+I.a~b~c~d+(I.Capacity)+I.z] + [CX+Cn] + [E±] ± [M1XM2XM3X(M.Capacity)+M.z]

Where **[I]** is information (first-order system structure data) and **[E]** is energy, **[M]** is matter and **[C]** depicts the aggregation of intrinsic and extrnisic communication trajectories among the system components and with an environment. While;

**[Cn]** depicts the intrinsic communication trajectories and mechanisms. **[Cx]** depicts the extrinsic communication trajectories and mechanisms, depicting the "structure determinism".

**[Hh]** is a coefficient expressive of an amount of uncertainty correlating the system's fragments ontologically, teleologically and spatiotemporally to an observer in relation to an observer – a relativity factor or second-order data.

The aggregation of [I] &[h] represent system creation/recreation data (ontology), while the aggregation of [Cn] and [Cx] in relation to [h] expresses the system's dynamics and a learning capacity (Maturana and Valera, 1928). It is what Stafford Beers (1972) describes as "the brain of the firm".

In living systems, new behavioural traits (a, b, c, etc.) are generally introduced to the system, in case of animals, through random mutation (producing correspondent decisive behavioural patterns), environmental factors and a limited portion of cognition, while in *Homo sapiens,* due to the capacity of observation, modification of behaviour occurs mainly through reason, cognition, social learning (imitation and reinforcement), education, and a limited portion of "social instincts" (i.e. natural dispositions).

(Darwin, 1871, pp. 77,100, 102)

This means that Real-living-systems are incapable of total reduction and cannot be expressed inclusively in mathematical formulas.

The concept of ontology (Corning, 1997; Gallopin, 2003), where life subsystems are subsumed within systems and systems are subsumed within suprasystems till a whole is reached, is very significant (a point extended in section -6) in scrutinizing both education and sustainability. A macro-organism is evidently the perceivable total of an entire world of micro-organisms cooperating constructively (beyond their perception) to finally function as a whole (Corning, 1997). By contrast, an aggregation of individual hive-bees constituting a bee-caste functions as a singularity, beyond the perception of its constituent members (Corning, 1997; Nowak et al, 2010), regardless of the Natural selection mechanism which has allowed such demeanor. In this respect, human beings are the unique species on earth who are evidently meant to survive through sophisticated social structures which they have no spontaneous access to; unless through learning, educating, planning, organizing, weighing preferences, creating and re-creating social agreements for cooperation, coexistence and governance (Maturana and, 1928; Corning, 1997). This might be due the existence of the Lateral frontal pole prefrontal cortex (Neubert FX, Oxford University, 2014) brain area, which allowed cybernetic skills to almost replace "instincts" (Darwin, 1871; Maturana and , 1928; Childe, 1936 cited in Corning, 1997) in relation to developing social behaviour and designing social interactions strategies which how morals might have evolved (Darwin, 1999).

Education may be defined as the built-in (intrinsic) communication system of our life organization, once such organization has been established through education. This is what Maurana and Valera (1928) define as an *autopoiesis* (Maturana and, 1928) and sets quality education with sustainable development in a circular causal relationship (Soliman, 2013).

## 4.1 CYBERNETICS LAYERS:

A Cybernetic perspective that concerns itself with the observed system and its dynamics, while disregarding the role of the observer is of a first order, and is the early form of understanding Cybernetics. While the second order cybernetic has extended this perspective to acknowledge and appreciate the role of the observer (the dynamics and purpose of the observer as part of the system) (Glasersfeld, 2002).

However, both first and second Cybernetics perspectives, tend to presume a totality of an observed system (or a focal system) as distinguished from a background environment (Maturana and Valera, 1928) and an indivisible totality of an observer's *role* (Umpleby, 2007) in spite of admitting there are several observers: "everything said is said by an observer to an observer" (Maturana and Valera, 1928), while it is empirically evident in relation to life equilibrium, that each human being is an observer/potential observer (Schwinger, 2004; Para-Luna 2009) with a different perspective while at the same time a subordinate actor within several social structures constituting a hypothetical whole of a life organization. Bridging the gap between the observational awareness of a subordinate actor within the system and the observational awareness of an observer; then again the observational awareness among the several observers involves an identity crisis.

Schwinger (2004) suggests that system dynamics and management cybernetics concepts are complementary to achieve a "necessary synergy" in managing social organizations, however, in the case of a human holistic life organization there is a prerequisite of creating such organization (Gordon Child, 1936, cited in Cornell, 1997). This is a coordination process among intrinsic communication, extrinsic communication and system creation information or purpose. [Cn] + [Cx] in relation to [h].

Expressed differently: as we all are entitled (and required) to create and maintain the economic and moral systems we functionally submit to (a point extended in section-6), we are the creator(s) of our own life "organization", when life itself is a given (Childe, 1936 cited in Cornell, 1997). In this case individuals would - through the stirred-up information ebb and flow, be oscillating among layers of (Cybernetic) observation, till the required initial unified vision towards creating and maintaining the organization has evolved then developed then enforced then re-evaluated and so on. A teacher in this respect has less authority than a system manager and more responsibility than a mere actor, yet in a sense, is both.

Education for sustainable development, on the other hand would be more than a stirred-up information ebb and flow [illustration-4]. It a double loop process where education and integrated management of resources interrelate to create and stabilize the life organization when an initial vision of sustainability has been created through education.

**Illustration 2: Quality education and Sustainable development**

Education for sustainable development may step in as a third order Cybernetics, as an integrative approach (Schwinger, 2004) of system dynamics, management cybernetics, and system creation.

## 5- ... AN OBSERVER

Von Foerster (cited in Ganville) also states the following:

First Order Cybernetics is the Cybernetics of observed systems
Second Order Cybernetics is the Cybernetics of observing systems.

Third order Cybernetics proposed by Robert Vale (cited in Glasersfeld, 2002) deals with human life organization as phenomenal! When man is *an* observer and man is the observed and when the unity of self-reference is achievable only through addressing the gap of awareness aiming at population self-organization, which in turn is achievable through self-awareness which in turn is achievable through bridging the gap between the levels of observation of an aggregation of observers.

Third order Cybernetics is Cybernetics of providence; the providence of the *Homo sapiens* over their own self-organization to achieve the internal operational "synergy" (Corning, 1997) called for, to maintain system equilibrium and thus to go in harmony with nature as a suprasystem.

A third order Cybernetic in the case of education for sustainable development may theoretically collapse into a second order cybernetics, when an actor becomes also an observer and a level of Univision (Corning, 1997; Schwinger, 2004) has been achieved.

At some point of my career development, when I wanted to share my vision on Cybernetics with some colleagues who were kind enough to be my "volunteer" students for an hour. I displayed a one-minute video for Stafford Beer explaining the control system of the Watt's Engine. I asked my students to write down the verbs they get to hear in the short film - in their exact tense:

My students came up with a list that included two verbs describing actions which were *not* performed by the system components. One was "operate" and the other was in the passive voice: "has been built".

This logically meant that the system - in spite of its autonomy - was not running totally on its own accord; some students proposed that maybe the system is actually controlled by the person "operating" it. The mind-game was a successful introduction to a rich discussion on cybernetic layers and the role of the observer. For decades, science has been founded on separating the observed system from the observer to comply with the fundamental "objectivity" principle (Forester, 1991) characterising scientific inquiry, but such objectivity in handling real-life systems has turned out to be unrealistic.

In the case of the Watt's Engine - as obvious in the language use and structure – autonomy occurs through provided energy and provided purpose constituting a whole of the machine in relation to a designer and an operator.

In the case of natural systems this "wholeness" representing the process of information and energy/matter through communication as perceived by observation can be inferred in the addition of Planck's (1949) constant (cited in Umpleby, 2004) indicating that the "electromagnetic energy is emitted in bundles or quanta"; the genome as it evolves across generations of an observed population of organisms where a current generation is treated by conservationists as a representation of their ancestors while represented in their progeny based on characteristics (Ruggiero et al.,1994); the structure of the gene with its DNA, RNAs and protein (Johnson &Raven, 2002) studied in its genotype-phenotype correlations (Encyclopedia Britannica); the human body with its endocrine systems (Ophardet, 2003). nervous systems and in social organisations. A macro-organism is evidently the perceivable total of an entire world of micro-organisms cooperating constructively (beyond their perception) to function as a whole (Corning, 1997) while an aggregation of individual hive-bees constituting a bee-caste seem to function as a singularity, beyond the perception of its constituent members (Corning, 1997; Nowak et al, 2010; West et al, 2006) and regardless of the Natural selection mechanism which allows such demeanour. That all in addition to "Nature" - in its Dawinean sense - as "the aggregation of natural laws" manifesting themselves through the natural elements and constructing the amazing variety of life systems; or what Wilson describes as "the forces of the natural selection"" (Wilson, 1998)

In studying biological systems, the role of the observer(s) is usually marginalised if not totally denied. First order cybernetics acknowledges information only as it reveals itself in matter/energy but overlooks the wholeness of the system, perhaps because such wholeness involves the uncertainty of handling information beyond the testable substance of the system: The information which defines the circularity, or the "cybernetics of cybernetics" (Foerster, 1979)

In machines, and for a system to operate efficiently, there must be a sort of harmony (compatibility) between the awareness of an observer operating or exploring the system dynamics and the awareness of *the* observer of the system who has conceptualised such alignment of matter and energy to produce such autonomy. This distinction is trivial in engineering but extremely important when cognition is in focus.

In case of machines; in order to address the gap of awareness, the potential operator of a given system must become also an observer through knowledge transmission, either via exploring cognitive artefacts (books, manuals, expressed designs, patents, etc.) (Jones and Nemeth, 2005) or a training. Once this awareness gap has been addressed, the distinction between observers vanishes and system sustainability is almost guaranteed.

While the autonomy of the living systems (organisms) and the spontaneity of animal social systems do not strictly involve such distinction between observers.

On the other hand organizations are basically created through social agreement (Garett Hardin, 1968; Schwaninger M., 2004). Once such mutual agreement comes into force, it becomes a defining maxim for a higher-identity system; operating through an hierarchy of agreed-upon processes and mechanisms towards coexistence, cooperation and governance. A higher-identity system once alive through autonomy, and as it still acts for the benefit of each of its constituent members, would hold (through decision making mechanisms)  the welfare of the group above the welfare of an individual in case of any conflict of interest. Hardin(1968) calls it a "mutual coercion".

At this point, addressing the gap of awareness (realizing synergy) becomes of the highest importance for the sustainability of such higher-identity organization (Stafford Beer), which is functionally able to mobilize more energy and matter into itself(ideally for the fairly-distributed energy/matter supply for its individual members), than the sum of the anticipated energy- and-matter supply each of its individual members- would separately get, due to the subsuming nature of the aggregated elements (forming the body of the organization), and the sophistication of communication facilitating the energy and matter import.(illustrations-2&6&12)

Nonetheless, the ebb and flow of information through communication trajectories, and the fact that the roles of an observer and an actor and *the* observer swap, interrelate, interact and intersect because of human rationale and the constant change of social settings, get education to assume a vital function in any social organization.

'Closed' network of people in interaction producing a whole.
Organisation emerges when members of a collective produce a closed network of recurrent interactions. Closed network, means that the collective has decision rules and mechanisms to make up their own minds about relevant issues, producing through their actions and decisions a whole, which

maintains a separate existence. An organisation as identity and structure. (Espejo, 2003)

Moving to an even higher organization, in order for system Cybernetics to apply, and for the mutual benefit of human beings, a life organization of such a description must be established, through the existence of *the* observer who would provide a conceptualization of the process of a global cooperation versus the current global challenges then through education to share an initial vision of a mutual potential benefit for such cooperation, which - once acceptable by an observer, may lead to mobilizing energy and matter for a globe-population organization, then through education to develop the vision into a mission and mechanisms then through education to update the vision, mission and mechanisms through system circularity.

But what is education?

## 6- CYBERNETICS OF FORMAL EDUCATION

An educator by any name (facilitator, trainer, tutor, guardian, etc.) is uniquely given the providence over a great portion of information ebb and flow in our systems. The teacher in his/her most traditional role will "instruct" the students and will " facilitate" learning in the most liberated role given by the community:

| TEACHER IDEAL ROLE | STUDENT IDEAL ROLE [based on learning outcomes (Bloom's Taxonomy)] | | |
|---|---|---|---|
| Teacher role | Learning Domains | | |
| [based on Gravells'(2003-2007) teaching-learning cycle and Tyler's model] | Cognitive [Bloom(1960)] | Affective [Krathwowl et al(1964)] | Psycho-motor [Dave(1975)] |
| Communicate | Communicate | Communicate | Communicate |
| Plan Activities | Know | Receive | Imitate |
| Organize: | Comprehend | Responding | Manipulate |
| Assess: learning outcomes | Apply | Valuing | Precision |
| Evaluate learning outcomes | Analyse | Organizing | Articulate |
| Reflect/research | Synthesis | Characterising | Naturalize |
| Re-plan: based on new inputs. | Evaluate | | |
| Communicate | Communicate | Communicate | Communicate |

# A Cybernetic Perspective of Education

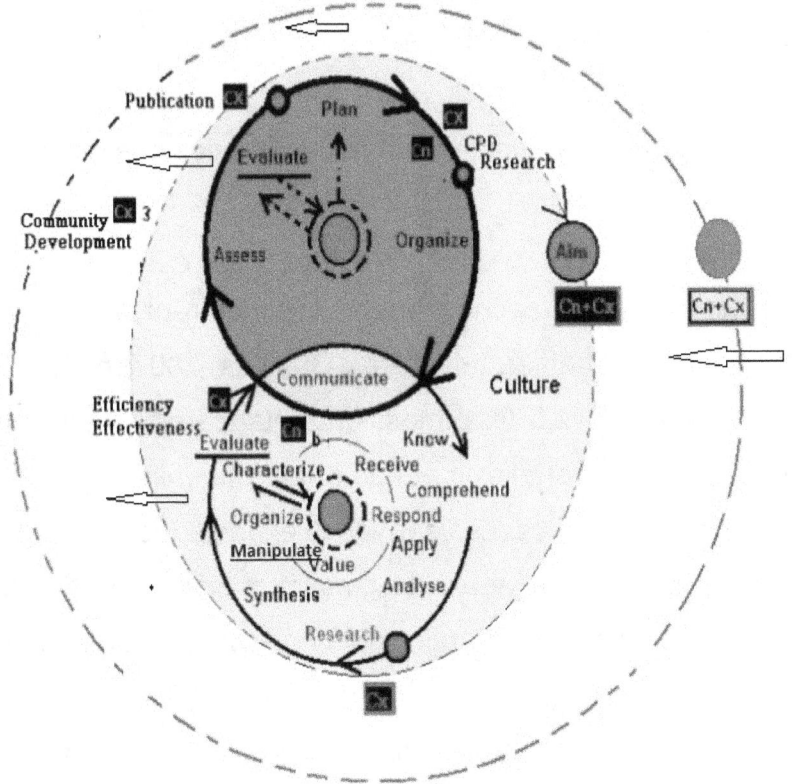

Based on Gravells (2003) Learning Cycle; Bloom's (1960) Taxonomy ; Tyler's (1949) Curriculum Model ; Maturana and Valera (1929) Structure determinism paradigm.

29

Gravells 2003; (Bloom's Taxonomy cited in Reece & Walker, 2007; Chen et al, Smith M.K., 1996)]

[

**Illustration 3: A perspective of the Cybernetics of education.**

Although the role of a teachers may have different names such as facilitator, instructor, trainer, tutor, etc. (Gravells, 2007, p11), research undertaken by LLUK( LLUK professional standards, 2009) indicates that all teachers undertake the same activities in relation to the teaching cycle and in all cases, learning is expected to occur through a controlled environment(at least to ensure safeguarding) (Gravells, 2007) and through the exploration of cognitive artefacts(Jones and Nemeth, 2005) (i.e. stored human knowledge); or through social learning(Bandura, 1971). A person who has assumed the role of facilitating learning and has agreed to undertake its responsibilities (Ingleby et al, 2010) will need to identify the learner's needs, select educational materials and study aids, assemble, organize and cascade cognitive artefacts or design learning activities in a certain manner, predictive of certain learning outcomes (cognitive, affective and/or psycho-motor), then assess learning and achievement of the students [illustration-4]. The language use is indicative of the difference in responsibilities. The role reverses during feedback (formative assessment) involving a new input to the process which is the total outcome of what has been communicated and/or what have been learned (Gravells, 2003; Tyler, 1949, cited in Villeneuve-Smith et al (1996); LLUK Professional Standards, 2009; Bloom's Taxonomy (1960)).

Such process is way too far from being natural or neutral regardless of the teaching strategies or techniques and regardless of the name of the role. It is the sum of an interaction and interrelation of each learner's purpose, needs and abilities with the educator's purpose, needs and abilities through the classroom culture. The term educator here refers to the tutor as a community mandate as shown in illustration (4). However, the purpose of the community is not communicated to the learner only through an educator, but through norms, laws, social modelling, social expectations and social pressures for self-effectiveness as expected to be leading to functional efficiency required for positive contribution to the community progress.

The purpose of the community gets communicated to the educator through the same culture of norms, laws, social modelling, social expectations, social pressures as well as professional pressures for functional effectiveness. Professionalism might get communicated to the tutor through teacher training, continuous professional development courses, quality assurance visits, observations, curriculum aims and objectives(if provided), professional standards, relevant laws and regulations, institutional policies and procedures, profession-related reading, own research and institutional policies and procedures (Gravells, 2007; Reece & Walker, 2007; IFL, 2009), in addition to the feedback from the learners (Gravells, 2007) and information from stakeholders such as guardians and governors (Houle's model of professionalism, cited in Tummons (2010). This means that before the teacher poses at the hypothetical starting point of planning at the teaching cycle, he/she has gone through a first round of research to familiarize with the social/professional requirements and expectations, initial teacher training (first step to), initial assessment (through communication with the learners) (Gravells, 2007).

An educator can never, in this respect, sit on the fence! By accepting their role of managing a classroom culture and enabling learning, they are held responsible for a portion of mass communication towards population self-organization of a given community. This community can be a small village or might extend over the globe; according to a vision! Functional skills as well as the ethics and social skills required for self-effectiveness, efficiency such as self-composure, resilience, teamwork, punctuality, creativity, reliability, celebrating diversity and perceptive tolerance can only be transmitted through social interaction and thus, in a way, through the classroom culture (LLUK. professional standards, 2009; Domain A)

Moreover, sharing information on nature, sustainability, and heritage are less likely to be shared naturally in every-day life situations.

Understanding education as a process of within-organization (family, nation, moral system, etc.) knowledge transmission, entails updating classical learning theories in light of the more recent ones to benefit from evidence-based research that must have been understood differently in a different environment with different contexts as well as different purposes.

Behaviorism, for example, as a movement in psychology started around 1913 by the publication of John Broadus "Psychology as the Behavior Views it" proposed that a conditioning experiment conducted on a lab dog (Pavlov's), was able to describe all what human behavior is about, and thus when applied to teaching was dependant on a presumptively more capable or knowledgeable person stimulating a change of behavior of less knowledgeable and capable individuals( i.e. students). That understanding was in light of the then available knowledge and within a certain historical context

Watson (1924) went as far as to consider that all behavioral variations among individuals are due to the learning experiences they are exposed to; and famously articulated:

Give me a dozen healthy infants, well formed, and my own specified world to bring them up in and I'll guarantee to take any one at random and train him to become any type of specialist I might select- doctor, lawyer, artist, merchant-chief and, yes, even beggar-man and thief, regardless of his talents, penchants, tendencies, abilities, vocations and the race of his ancestors" (Watson, 1924)

(What is Behaviorism; About.com Psychology Website)

Obviously no one gave him any infants, healthy or otherwise, to run his extremely specific experiment and the movement was later criticized vigorously by several critiques such as Myer (1988), Piaget(1996),Vygotsky (1962), Glasersfeld (1980) and others who saw the necessity to move on to a more cognitive /constructive approach(Peel 2005). Both Behaviorism and cognitivist were based on the assumption that teachers knew exactly what the students should learn and just needed to find out how to enable such learning (Gray, 2011)Watson's requirements for a successful conditioning experience of human behavior is a clear indicator that managing students' behavior through conditioning is near impossible, yet, Reece & Walker(2007) and Wallace (2005) thought that Behaviorism is still useful if used to plan reinforcement strategies of the desired behavior, in correlation with other theories. Using praise and delayed praise is a sort of conditioning of a humanistic approach (depending on purpose) in correlation with constructivism as it indirectly informs the students of certain social expectations. Also the association of stimuli used in the classroom and the alignment of resources and activities may also be predictive of certain cognitive outcomes, especially if the student is prompted to research and give an interactive feedback. If after interaction and research the feedback on the learning outcomes were consistent with the tutor's objectives then conditioning has worked as a "persuasive association" of resources and activities which actually what scientific inquiry and creative writing are all about. If not then an update has occurred and a new interactive input may, constructively, be taken into the process, enriching it.

Drawing on my own experience as a learner, I have attended a session on empathy and the tutor displayed a video on people moving around as would normally occur in life but the movie labeled each person moving around with a brief of a personal concern; such as a woman walking around with a: "19 year old boy on life support" visible label and so on. That short video was (successfully) meant to stimulate empathy and would probably get you to think twice before judging a person with an absurd appearance. That in relation to the rest of the session activities was transformational because it provided through the session structure and alignment of resources and through the well-designed activities an authentic experience I would not have gone through otherwise, and would have otherwise missed!

The rest of the activities allowed sharing experiences and perception of the attendants and went in conformity with my own research at my own time. Illustration (5) depicts a point where the learner may (and ideally should) research to validate learning outcomes, and then give feedback into the learning culture.

At another experience as a trainee teacher, I wanted to teach a group of volunteer students about classroom management. I decided to use an association of images in a certain alignment in a short self-made movie to depict the cognitive value of the classroom culture, then I opened a discussion to discuss the experience, then I shared the resources for a delayed feedback. The experience seemed to me through the feedback from learners and mentor as a success.

Surprisingly, a teacher cannot avoid conditioning, and can only select to use it positively or negatively in relation to the cognitive-affective aims and objectives. Darwin believed that conditioning and reason are the two faculties that have probably caused our moral evolution (Darwin, 1871) to occur.

A teacher who ignores praising a student for an achievement might be reinforcing passiveness and one who comes late for the session or fail to show compassion or empathy(Reece & Walker, 2007) within the professional boundaries (Gravells, 2007) or who disregards swearing, disruptive behavior or bullying during the learning times is sending strong conditioning messages due to the power of reinforcement resulting from the extensiveness of social contact and interaction, whether intentionally or subconsciously.

Conditioning is what marketing and social learning are all about. It is all over the place. Conditioning is practically forging our economies and social relationships (Mariotte, OIKOS) together with the faculty of reason (Darwin, 1871).

In the recent past, the funeral of the deceased African leader Nelson Mandela (2014) might serve as an example of a global positive reinforcement of an altruistic *life pattern,* whether agreed with his politics or not.

In a historical perspective, it was generally assumed before 1950, that all people learn the same way and therefore all learning activities where known as "Pedagogy" which literally translates (from Greek):"leading the child"(Holmes and Abington-Cooper, 2000). Later on, Knowles (1970) suggested a significant change in the way education for adults is designed and delivered, and so he became known as the father of Andragogy (the art of teaching adults). Some think that a new era has come for the Heutagogy (Holmes and

Abington-Cooper, 2000) or the self-directed learning which looks like a noble cause, however, on the practical side of it and for a pure self-directed learning experience to occur, a learner must go to a purely natural learning environment where there are no posters or multimedia influences and must not read persuasive texts or be exposed to social media or marketing channels (Mariotte, OIKOS) and must not be exposed to his/her peers and parents' inconspicuous, conditioning!

On the other extreme of behaviorism, comes the Humanistic education which regards learning as a "self-fulfillment" (Rogers, 1951 cited in Harp & Raw)  experience and therefore calls for the prevalence of self-directed learning "(Patterson, 1977), and for "liberating students'" (Gray, 2011). The movement was influenced by the views of Roger (1951) Knowles (known as the father of Andragogy) and Maslow (Gray, 2011).

Abraham Maslow's hierarchy of human needs (1943), proposes that the needs of human beings may be listed in five hierarchical levels; physiological, safety and security, love and belonging, self-esteem and self-actualization.

Yet Abraham's hierarchy of needs seen in a Cybernetics perspective and in light of other paradigms such as the structural determinism (Maturana and Valera's, 1928), transforms spontaneously into a cycle of needs; and this would be in reply of one simple question: Given that food and water are of a first priority among all human needs, where would the food (or water) come from?

The answer would be that human organismic needs are obtained through a network of relations, interrelations and interactions that set the need for love/belonging as well self-effectiveness (realistic efficacy or self-esteem) on the way to achieve the basic needs. A newly born baby is fed, maintained and kept clean through love and belonging and a church providing a meal for a destitute does this based on love of humanity, even if the official responsible for the giving process is passionless. Normally an adult needs to demonstrate efficiency (which is also "self-actualization" or realistic "self-fulfillment") to take a job and thus to eat and reproduce; to take a job, one needs to provide at least two references and a qualification, obtained through self-effectiveness and efficiency.

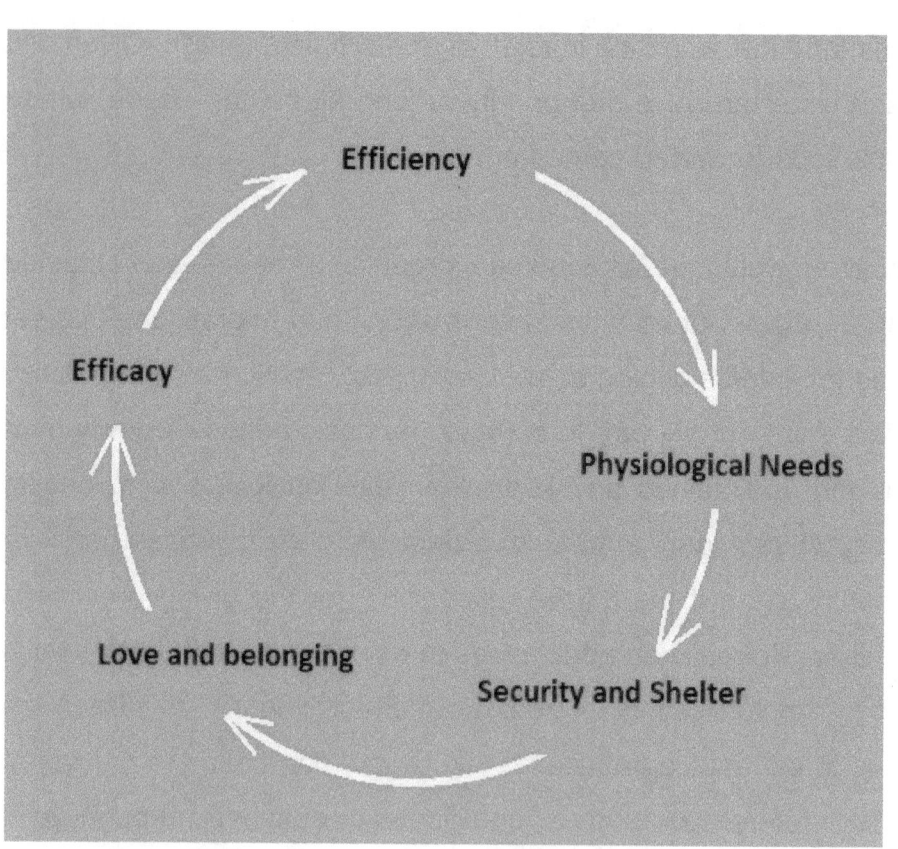

**Illustration 4: The Cycle of Human needs Based on Maslow's (1948) hierarchy of human needs and the structural determinism principle of Maturana and Valera (1928)**

**Efficacy** involves a realistic self-esteem **while efficiency** involves a realistic "self-fulfilment".

When we meet a person in need, we may assume that his/her cycle of needs fulfillment has swirled several times at several levels before he/she is in need of food (again) and by helping this person we are expressing compassion and sympathy to begin with.

In case of dependants and oppressed, individual might find it easier to trade their self-esteem-tied up to self-effectiveness in return of the basic needs including the need for security (Friere, 1993) and then pseudo-self-denial might become a way of self-effectiveness. It is a cybernetic cycle of needs that spirals through achieving levels of self-actualization or what Roger (1951; cited in Harp & Raw, 1977) comprehensively describes as "self-fulfilment".

Retrospectively, while Behaviorism is focused on stimuli- response processes and behavior control, and Humanistic education is based on the ambiguity of self-fulfillment and the controversy of liberating cognition from social pressures, Cognitivists focus on how the students acquire, organize and evaluate knowledge, which makes it of the highest relevance to the current educational systems and widely embraced as a genuine educational movement. However, the theory is based on a general assumption that teachers knew what the students ought to learn (Gary, 2011).

The Bloom's Taxonomy (1951) cited in Reece & Walker (2003), defines the domains of learning into Psycho-motor, Affective and Cognitive and defines the degrees of difficulty in learning within the cognitive domain as: Knowledge; comprehension, application, analysis, synthesis and evaluation (Reece &Walker, 2003). The affective and psychomotor domains were added later to the taxonomy by Krathwowl (1964) and Dave(1975) in correspondent order. However, in avoidance of behaviorism, the approach overlooks the cognitive value of the classroom environment and thus classroom management has often been dealt with, by educators, as an issue of handling challenging behaviour.

However if a purpose of learning is provided, the challenging behavior may ideally be addressed as an integrative component of the educational process, and thus handled through multiple teaching techniques to make the session interesting and relevant to the learners(preventive approach), and through the positive reinforcement, as well as using endorsed disciplinary measures in case of extreme disruption, to ensure safeguarding and learner-learner respect then also through empathy (one-to-one counseling). That is an integrative application (Peltier, 2001, p.47 cited in Reece & Walker, 2007; Peel, 2005) of Constructivism, Behaviorism, managerial skills and humanistic education.

Models of communications are critical to systems. The transactional analysis is one of important communication concepts, developed by Eric Berne (1964) in regards of one-to-one dialogue proposing that each one of us has three levels of inner identities or egos; the child, the adult and the parent, which are all present but one may dominate over the others in different situations causing an unbalanced conversation, unless it is flowing in an adult to adult mode of communication. A teacher may consider not to be dragged into an adult-child or a parent-child mode of communication with learners and rather to always try to keep the conversation logical and non-judgmental. (Appleyard & Appleyard, 2010). Additionally, communication barriers such as the psychological, semantic or the physical barriers described in (Appleyard & Appleyard, 2010) need to be taken into consideration. Communication is what makes any system, a system! And therefore very significant in the educational process.

Moreover, assessment is another essential component of the educational process (Gravells, 2007) as it serves in general terms to "determine the importance, size or value of"(Longman, 1984:86) things and processes in general terms and more specifically in education. It aims at replying the question in Jarvis (2005): How do we know if that a person has learnt something?", and that is in regards of function, while in regards of application, strategies and techniques of assessment vary according to their purpose and power (Higgins, 2006).

Eventually, there are two kinds of assessment: Formative and summative. The former is conducted to assess the learning process as it goes on and the latter is to satisfy the needs of the society mainly for the "certification purpose". The former is an in-going process conducted in several techniques such as asking questions during the activities to diagnose learning needs/difficulties or through feedback during activities and projects, while the latter is conducted at the end of the course, in various techniques such as exams, essays or projects, aiming at evaluating the overall learning process (Reece and Walker, 2007). Testing as a specifically summative assessment tool, functions effectively as it measures the achievement of the students according to norms and criteria. The evaluation of the results are either "criterion referencing", which is only effective if the marking criteria are well defined (Walkin, 2000), or "norm-referencing" as in national tests when results are compared and moderated against certain norms across generations, such as the bell curve, to ensure equal opportunities (Walkin, 2000). According to Wallace (2005), planning accurate assessment depends on the validity, reliability and sufficiency of the assessment techniques: An assessment tool should be valid in the sense that it is capable of assessing what it is supposed to be assessing, and reliable in the sense that it is likely to produce the same results if marked by two or more different teachers, in addition to having to be also sufficient in providing adequate evidence of learning and achievement. Additionally, assessment techniques must also be inclusive to ensure "equality of opportunity" which would mean for example, to plan alternative assessment methods that are more convenient to one or some individuals, yet as valid and reliable as the replaced ones (Wallace, 2005). Finally that it must possess the power of discrimination in the sense that it provides results allowing the discrimination between the strong achievers and the poor ones (Reece and Walker, 2007), basically to satisfy the needs of all the stakeholders in the field.

From the overall discourse in section till section (6), we may define education in a Cybernetic perspective as the within-organization communication trajectory (Cn) that needs to be informed by the purpose of the organization and developed through the feed-forward-feedback trajectories to continually update the vision and mission statements driving the system's mechanisms. The system in turn communicates outwards - ideally, forms of developmental solutions and practices to contribute to the progress of the community. The successful coordination between (Cn) and (Cx) in relation to (h) represents a learning capacity and a management capacity or what Beer call "The brain of the firm".

## 7- NON-FORMAL EDUCATION

Learning occurs. Whether consciously or subconsciously. Laws of nature are constantly sending cognitive messages through things, beings, domains, times and spaces (Mataurana and Valera, 1928). Due to the mental faculties of human beings, human cognition and modification of behavior are an integration of reason, social learning(imitation and reinforcement) and a relatively(compared to other beings) limited portion of "social instincts"(i.e. natural dispositions).This means that every single gesture and every single word we write, say or refrain from saying, is likely to be a cognitive message to someone somewhere around us.

People adopt ideas and embrace values when such ideas and values relate to their logic and are able to comfortably fit into their knowledge-reservoir structure (Glasersfeld,2002). That makes education a double loop process of persuasion - adoption, logical proposal- rational acceptance, seeking information-giving information, implying-embracing, expressing-comprehending, stimulating-responding, etc. (illustration-4). We teach and learn, then, all the time and in all life domains; whether consciously or subconsciously (Darwin, 1871, pp. 77,100 102; Mariotte, OIKOS; Mataura et Valera, 1928)

## 8- EDUCATION FOR SUSTAINABLE DEVELOPMENT: THE INTELLECTUAL VIABILITY

"...great philosophers and discoverers in science, aid the progress of mankind in a far higher degree by their works than by leaving a numerous progeny (Darwin, 1897, p106)

## 7.1 INTRODUCTION

It might be stating the obvious that the "survival of the fittest" used by many as a slogan of legitimate selfishness, does not, and never has handled the fitness of an individual unless interrelating to the fitness of a relevant population over a period of time exceeding the life span of the within-population individuals. That should explains why the evolutionary perspective of fitness is always related to reproduction and accumulation of traits.

Expressed differently, that the natural selection acts on the advantageous traits of an individual as long as they provide a potential inclusive fitness for a population of "organic beings"(Darwin, 1999). Within-population Individuals have relatively short life span compared to the population as a life organization (higher identity organization) and while the natural selection acts slowly but steadily - hand in hand with other laws of nature, on the phylogenetic traits of populations, it inevitably leads to the success (and advancement) of some species and the extinction of others; what Darwin calls ("divergence of character").

Although natural selection can act only through and for the good of each being....3
.. yet of those which do survive, the best adapted individuals, supposing that there is any variability in a favorable direction, will

tend to propagate their kind in larger numbers than the less well adapted.

If we underline the words "each being" and "individuals" and then "their kind", the frame of reference will confirm the suggestion that Darwin introduced "individuals" as representing a phylogeny.

Another example of such frame of reference.

Preservation and accumulation of variations, which are beneficial under the organic and inorganic conditions to which each creature is exposed at all periods of life. The ultimate result is that each creature tends to become more and more improved in relation to its conditions.

The word "creature" here, in light of the language use and word sequence is the aggregation of:

Firstly ,a modified organic structure of a population seen as a singularity (genome), which is evident in the combination of the expressions "accumulation of variations" and "ultimate result" and "creature") and secondly that that modified organic structure is a representation of a population of progenitors who shared identical traits and where exposed to trait modification across generations, till they were represented by the current population (evident in pairing the expression of "each creature" with "periods of life", accumulation" and ultimate results" then the use of "each creature" again.

That makes a living individual, according to Darwin, a representation of a current population that is in turn a representation of all within-species individuals who have ever existed on earth.

This makes the act of reproduction itself, surprisingly, altruistic. A sea turtle (Turtle World) who takes a (relatively) exhausting trip from the ocean to the beach and digs a ditch where she can dissipate life substance out of her body into the earth to potentially contribute to the inclusive fitness of her species, leaving then the offspring to their destiny where only 2% of the youngsters may survive, might be as altruistic to her kind and to life on earth as the hive-bee who serves the productive fitness of the queen, in spite of being sterile herself.

Mathematically and causatively, they might be equal in contributing to the preservation of the existence of their kind although comparative statistics in this regards are not available - to the best of my knowledge. By all means, they both have incurred a cost that they have not and are not likely to be directly compensated for, unless we are discussing a life after death. Both would be dissipating hard-earned energy out of themselves into the environment as an act of spontaneous kindness towards their species/population/kind. In this sense the within-group competition might be regarded as relatively superfluous, especially if seen (from one perspective) as an endeavor to eliminate the maladapted individuals, to enhance the phylogenetic product of the group (Darwin, 1999, p. 103), which occurs in our social systems ideally through justice systems (Darwin, 1999) and governance, due to the nature of our cybernetic viability (illustrations 12 in relation to illustration 2& 6).

However, once reproduction has been conceived as a selfish endeavour of survival, biological altruism must become overwhelming (Rachlin, 2002) and so it was for generations. (Rachlin, 2002; Huffington Post: David Sloan Wilson, Richard Dawkins, Edward E.O Wilson and the consensus of many).

## 7.2 THE CRISIS OF COOPERATION

The success of any species is tied up to discriminative kindness. West et al (2006), explains an interesting experiment conducted on micro-organisms communicating for the purpose of cooperation necessary to perform several essential multicellular processes such as nutrient acquisition and dispersal, depicting the generic crisis of altruism or cooperation *in nature*. Some individuals in this bacteria groups namely *Pseudomonas aeruginoshe* produce a "common good" for the benefit of the group (sidrophores scavenging iron in this example), however, such production might be open to plunder by "cheats" within the same group who do not produce such organic product and who would use the plundered sidrophores to out-number the altruistic productive individuals. When selfish individuals ("cheats") out-number the altruistic one beyond the capacity of the patch/colony to provide nutrition and other essential organic products, the group inevitably perishes. West et al (2006), then explain how the altruistic bacteria uses two mechanisms to insure the sidrophores will be utilized exclusively by relatives, who according to the Kin selection theory, must be carrying the same altruistic traits and are likely to keep producing beneficial products for the survival of the colony, which may in turn, result in a potential inclusive fitness for the whole group. Those two mechanisms are the "limited dispersal", and "kin discrimination" (the repression of competition).

The Kin recognition and kin discrimination in this example occur beyond the perception of the individuals while in higher organisms occur consciously, through the sensory system (based on physical characteristics such as the smell, features or location, as relatives tend to live in the same vicinity) (West et al, 2006).

The success of kin discrimination would *naturally* lead to the growth in number of the connected group and a "gradual advancement of the organisation" (Darwin, 1999) leading in turn to more growth in population (inclusive fitness).

Park (2007) describes the potential inclusive fitness of species resulting from the Kin selection as occurring through three integrative processes: The cumulative impact of a gene on the individuals and their kin, (possessors of the same gene) and the Kin selection; which is the natural selection of the genes with encoded altruistic traits in addition to the kin recognition as a mechanism of spotting the relatives candidates for altruistic actions (Park, 2007).

Explained more plainly:

The basic idea of kin selection (a level of inclusive selection) is simple. Imagine a gene which causes its bearer to behave altruistically towards other organisms, e.g. by sharing food with them. Organisms without the gene are selfish—they keep all their food for themselves, and sometimes get hand-outs from the altruists. Clearly the altruists will be at a fitness disadvantage, so we should expect the altruistic gene to be eliminated from the population. However, suppose that altruists are discriminating in who they share food with. They do not share with just anybody, but only with their relatives. This immediately changes things. For relatives are genetically similar—they share genes with one another. So when an organism carrying the altruistic gene shares his food, there is a certain probability that the recipients of the food will also carry copies of that gene. (How probable depends on how closely related they are.) This means that the altruistic gene can in principle spread by natural selection.

So the overall effect of the behaviour may be to increase the number of copies of the altruistic gene found in the next generation and thus the incidence of the altruistic behaviour itself. (Stanford Encyclopaedia of Philosophy, 2003)

Regardless of the recognition and discrimination mechanisms, the inclusive fitness resulting in species success may be regarded as the ultimate result of perpetuating discriminative helping-tendency traits when cooperation *happens to be* of a critical inclusive advantage to the population (as in the example of the *Pseudomonas aeruginoshe),* or when the benefit of cooperation "outweighs" its cost leading to a natural selection of such traits (Cornell, 1974; Vareses, 2001; Doebeli and Hauert, 2005). In humans, the relevance and significance of kin discrimination to trait-transmission and cooperation seems to deminuate as we advance in civilisation (Darwin, 1871).

However, the selection paradigm does not shed much light on the cognitive side of the evolution story, when Kin discrimination is not the case and when human rationale is not involved, how does the "weighing" phenomenon occur and what mathematical-cognitive-reasoning-planning skills allow the navigation of the environment, measuring probabilities against *preference* and making a decision towards the progenitor altruistic behaviour, given that an initial decision to go through the first experience must precede any incident of reinforcement or mutual benefit.

In other word, in cases other than direct kin nurture there is the difficulty of deciding whether a specific action is altruistic unless related to several situational and contextual factors and sometime a complexity of a social structure and elements of time and space (Rachlin, 2002). Moreover, when altruism as explained in terms of mutual benefit occurs, a distribution of role must be involved and thus a question on which role gets selected to be stabilized arises and if more than one role is stabilized, what kind of rationale is used to distribute such roles among the group members and in which proportions? It seems that an altruistic behaviour, in any case other than helping a kin to reproduce, must evolve under a *providence* or a social agreement - which is usually beyond the perception and cognitive abilities of the groups involved - or as a foolish behaviour of a suicidal individual (Wilson, 1975, page 578 cited in Rachlin, 2002) risking their own fitness (and subsequentlly the inclusive fitness of their species) to help or increase the fitness of others who may or may not benefit from the emergent behaviour in different settings, especially if the first experience (as out-of-norm) is likely to produce a negative reinforcement (Darwin, 1999, pp. 101,102), given that altruism is a tendency to help rather than a code of conduct (Vareses, 2001).

In case of self-regulatory machines, behaviour or "control" has been "built into" the alignment of matter (Stafford Beer in explanation of the viable system, Youtube). Each actor in the system acts according to a 'role given" by an observer (Glasersfeld, 2002) and therefore cooperation is mandatory.

In case of animals, random mutatin are said to be responsible for the emergence of new behaviour and thus when altruistic traits are perpetuated in a population, individuals of the population will amazingly (Ridly, 1995) have the spontaneous skill to act on those tendencies with no need for training or education or a manual and regardless of situational factors.

## 7.3 A HIGHER ORGANIZTION: hICEM.

In case of human beings, the relevance and significance of kin discrimination to trait-transmission and cooperation deminuate as we advance in civilisation. Through the observation capacity and intellectual data transmission (language, models, education, etc.) morals have "evolved" at a certain point of history separating gradually such social tendencies from the genetic realm (assuming they have been once there), that the behavioural traits of kindness have become transmittable mainly through culture even in dealing with biological offspring. Actually, a great deal of human-life system information are now "attached" to our *phylogeny* through coding and the use of cognitive artefacts rather than into our genes. Such codes by adoption are capable of manipulating natural fundamentals so as to mobilize more matter and energy into our physical existence which has now encompassed all the natural elements and species through the capacity of "steermanship". What is wings to a bird compared to one preserved design of an aeroplane to humans.

That is the viability of our species has become rather intellectua (Vareses, 2001) as we strive to achieve the required "advancement o" the organization" (Darwin, 1999). Isn't it strange that we use the word "adopt" to indicate fostering ideas as with children.

Scientists have found that one reason life exists is because the amount of matter supersedes the amount of anti-matter in the universe [anti-matter annihilates matter releasing energy] (NASA, 2008). But death as we try to avoid, is not the absence of life and is not equal to nothingness (yet); it is simply the prevalence of "lower-level" of life organizations over higher-level ones, leading to the decomposition of an organic system into more simple units that would no longer be connected Cybernetically, under the same old  higher identity (hICEM). All organic beings whether consciously or subconsciously strive to maintain this higher identity long enough to replicate their existence by reproducing "more offspring than can survive" (the struggle of existence) and thus get even "higher in the scale" (Darwin, 1999) resulting in the continuation (renewal) of life and an arguable enhancement of species, in spite of the relatively short life span of the individual life systems. It is what Corning (1997) describes as the survival of "*genome*". But the human genome is much more complex than its biological existence as it relates to an intensive cultural heritage and natures fundamentals through cybernetics. It is who we are here-and-now stamped by the past and consciously reaching out for a future.

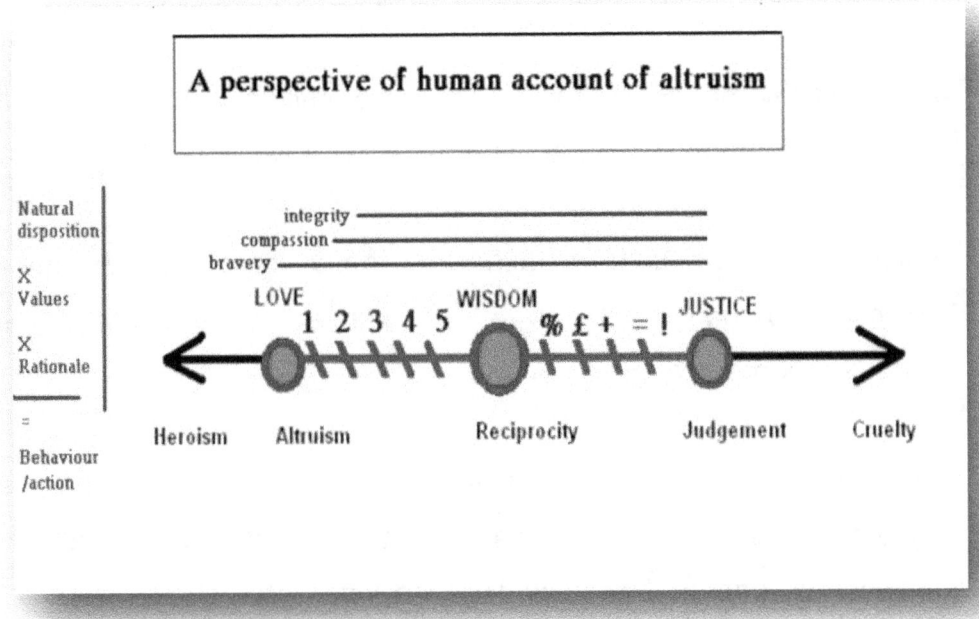

**Illustration 5: A perspective of human altruism.**

The "survival of the fittest" is thus in itself inclusive.

By this observational revelation, we find ourselves compelled to dig into the purposes of nature to design our own activities and work out their consequences, when a standardized form of altruism, may be rather suicidal to human individuals and groups. Cooperation in our world of cybernetics, as much as it is potentially beneficial, must translate into perceptive organizational maxims that takes the intergenerational prosperity of humanity seriously into its highest account.

A potentially shocking translation for Hamilton's rule in its non-mathematical form in a world of human morals would read:

Love (as the core value of altruism), can be spread by the natural selection so long as justice − (the ultimate value underlying Kin discrimination process) prevails.

An immediate reaction to this statement would be that there are several levels of altruism among which cooperation is one and that cooperation does not necessarily need love as much as it needs an awareness of a mutual benefit, however, it is such association of values that produces the balance underlying ethics of constructive cooperation for achieving a mutual benefit. Human cooperation devoid of the love component is *temporary;* inevitably dilapidating into ultimate conspiracy, dispute, and wars, once or even prior to mission is accomplishment. This may be illustrated in the cooperation of cancerous cells to lower the fitness of the healthy cells (those cooperating constructively), then once the population of the former has reached a certain limit of prevalence or have managed to hit a control spot, it lead to the decomposition of the whole system/body into "lower levels" of existence (death) including the decomposition of its own mass.

Another limitation to this interpretation would be that love and justice are ultimate values while altruistic behaviour and fair play are situational, which is indisputable, however, human altruism *is* an ultimate value; a tendency to give or help, with no genetic clue on how to act on such tendency (Vareses, 2001).

Levels of human altruism would begin with helping kin and through reciprocity, to a tendency to die for the beloved ones (Stanford Encyclopaedia of Philosophy, 2003). Of course, natural dispositions have no right or wrong, they just exist. However, "reason" interferes sharply to restrict or "extend" (Darwin, 1971) such natural tendencies. The more we realize the value of communication in passing such moral association of love and justice, the more altruism would expand beyond localities, borders or blood relation leading to constructive global cooperation and potential species success.

As man advances in civilization, and small tribes are united into larger communities, the *simplest* _reason_ would tell each individual that he ought to extend his social instincts and sympathies to all the members of the same nation, though personally unknown to him. This point being once reached, there is only an artificial barrier to prevent his sympathies extending to the men of all nations and races

(Darwin, 1871)


## 7.4 SYSTEM MAINTAINANCE: SUSTAINABILITY

Unfortunately, because of the same super intellectual capacity of our species, selfish individuals may gain ultra-power through sophisticated communication media and manipulation of nature's fundamentals in a sort of "pseudo- altruism" or temporary cooperation. Selfish behaviour that does not extend to the rest of humanity can thus be individuals or among a group. Such behaviour was natural at a certain level of our evolution. As the humanity advances by reason towards promoting global citizenship and no-discrimination trends, before actually handling world governance issues to ensure the protection of the kind/tolerant individuals and groups, kindness can get fatal to our species as observed in all other life systems. Selfish entities whether individuals or group may use a dual communication process (What Darwin in his *The Descent of Man* accounts as "lying to the enemy") to access and maintain control thus prosper and then ultimately collapse for the lack of real cooperation and destroy humanity. One selfish influential decision, indifferent of life, in a world of mass-destruction weapons can have an irreversible impact on our survival. That is to suggest that our systems of justice and our concepts of tolerance must be revised in light of their survival values rather than their epics or metaphysics.

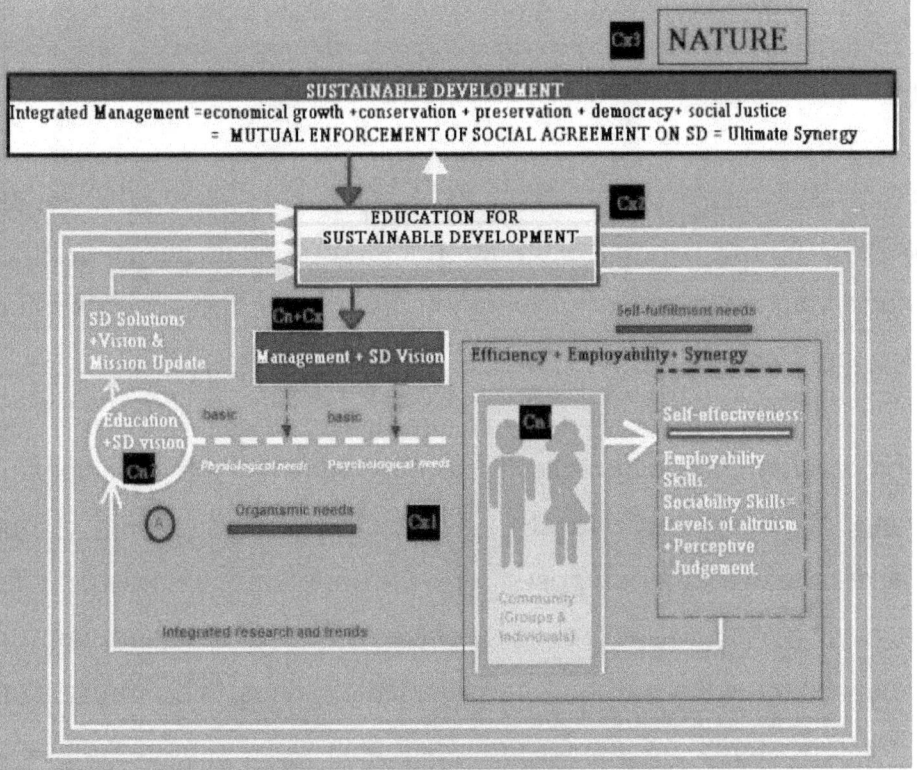

In regards of the need for global governance in regards of development and preservation of natural resources, it is an overwhelming paradox to depict the harmful impact of "human activities" on the environment as of a product of a unity (Gallopin, 2003) while considering the solutions, as individuals, moral systems, economic systems, ethnicities, communities and nations; knowing that such systems do not function in conformity, with some actually invasive of others.

Illustration 6: A Cybernetic evolutionary perspective of education for sustainable development (Maslow, 1943; Rogers, 1951 cited in Harp & Raw, 1977); Sustainable development indicators, 2013; The Millennium Declaration, 2000))

The real challenge of preservation and sustainable development is not in the shortage of natural resources (although it finally leads to such shortage), but in the lack of cooperation (point extended and illustrated in sections 8.2& 8.3).

Having reached this point of discourse, I would revisit the definitions of learning, cognition and education as ubiquitous components of our inclusive fitness and as proposed, one-of-two integrative mechanisms for achieving sustainable development and as the potential intrinsic communication system of our life organization once this organization has been founded on education. This sets quality education and sustainable development in a circular causal relationship.

Learning is the self-directed responsiveness to any change of information communicated to a given system. In this respect Maturana and Valera (1928) have suggested that learning (not necessarily cognition) is a deterministic state of any system as it responds (on the least) to the natural forces; some observable examples of this are the solar system and the dust particles as they sit on a mountain or interacting within a tornado's whirl, among other aspects of nature. Learning occurs spontaneously through coming into contact with other systems via intrinsic and extrinsic communication trajectories that are detectable through the observation of relationships among the system components and its external interactions, yet not always easy to code. Example of such is the cell-cell communication (Raven & Johnson, 2002) the nerve-cell communication (BBC Science & Nature).

Education, on the other hand, is a deliberate human intervention to influence the flow of information packets driving our life systems ideally towards a noble goal or a set of goals, based on the unique ability of observation and strategic planning (i.e. Cybernetics) and the sophistication of human communication systems. This happens mainly but not exclusively through the use of language.

Culture is constructed with language that is productive, comprising arbitrary words and symbols invented purely to convey information. In this respect Homo *sapiens* is unique.

Animals have communication systems that are sometimes impressively sophisticated, but they neither invent them nor teach them to others (Wilson, 1999).

This phenomenon is what might have caused or has been caused by a shift in our genetic structure that sociality traits have become transmittable through culture rather than reproduction.

By all means, education may in cybernetic evolutionary perspective be defined as the within-organization communication trajectory of the human species or the intellectual reproductive system of humanity.

Education for sustainable development does not teach answers but asks for ones. It creates a positive culture for creativity, problem solving and resilience - self-control, efficacy and efficiency with the association of love and justice as a moral coefficient of all activities and with a shared vision towards the higher good of preserving humanity and nature.

It would be logical to assume that there are already economic, political and even ethnic systems mobilizing power to access full control as obvious from the global challenges and someone needs to pick up the phone on behalf of humanity through social initial agreement among a group of decision makers or "intellectually fit" individuals who can influence the morals of populations through mutually-persuasive communication, promotion of relevant scientific knowledge, publication, social activism and tutoring then pave the way for democratic global governance. Ethics need to be re-set in relation to their species-survival values, which is only achievable when there is an initial unity of self-reference as a human species, which in turn is achievable through education for sustainable development, which in turn is achievable through integrated management of resources and social justice, which again is achievable through narrowing (hypothetically bridging) the gaps between our observation layers through education and anew (illustratin-8).

Activism, needs to begin at a certain level of awareness which depends on a level of scientific objectivity with a level of acceptable uncertainty of an observer to reach an initial level of cooperation capable of construction and mobilization then spirals in its growth through levels of synergy and awareness till it hypothetically reaches a point of harmony with the whole system of life: nature.

The communication of values is inevitable. A message totally devoid of value is communicative of void. A chance for creation missed! And if "no-thingness" in the physical world has a power to annihilate things (matter-antimatter relationship), how would the absence of constructive message which could have communicated a value of survival.

# 8- CYBERNETIC SKETCHES

## 8.1 THE ARAB SPRING AS AN ICEM

If the Arab spring phenomenon may be assimilated is an example of how social systems with not adequate defining information may come to.chaos.

The Arab spring known as the call for "*Esqat Alnezam*" which literally translates to: "tipping the "system" (regime) over", Due to the compatibility of information transmitted through the sophistication of the social media communication systems, it evolved as a tornado of anger, however, as it had no development (illustration 6 &13) defining information, it quickly dissolved in some geographical regions and was subsumed into more sophisticated systems in other regions.

A packet of information thrown into an environment ready for population-self-organization into an hICEM, under convenient social "pressure and temperature" with no vision on how to lead the future.

I have seen Egypt under the military rule (again) as an alternative for a much worse scenario of chaos. I've seen a video circulating on the social media from Syria of people dissipating life out of two men in apparent pleasure and a video of two children: *Children* made to sit on the ground and hear their death sentence by an adult in a black mask then eventually get shot! I saw terror on the faces of the two children and submission to adults who should have provided love and care. I've also seen reproach!

As an educator, I wish I could invite the inventor of the hierarchy of human needs (Maslow, 1943) to come and check if those children needed food more than they need love and belonging to survive!

Who picks up the phone on behalf of humanity?
If chaos is what we believe life is all about, then chaos is all what we get.

## 8.2 EDUCATION OF GENDER EQUALITY

Analysis of the world hunger statistics shows that the number of hungry people in the world could be reduced by up to 150 million, "If women farmers had the same access to resources as men", according to *Women in Agriculture: Closing the Gender Gap for Development*, FAO, 2011 report, and knowing that much of today's hunger is caused by wars, civil wars, destruction of infrastructure then hunger traps, which are related to social issues rather than natural disasters or shortage in resources, it may be about time we try harder on gender issue.

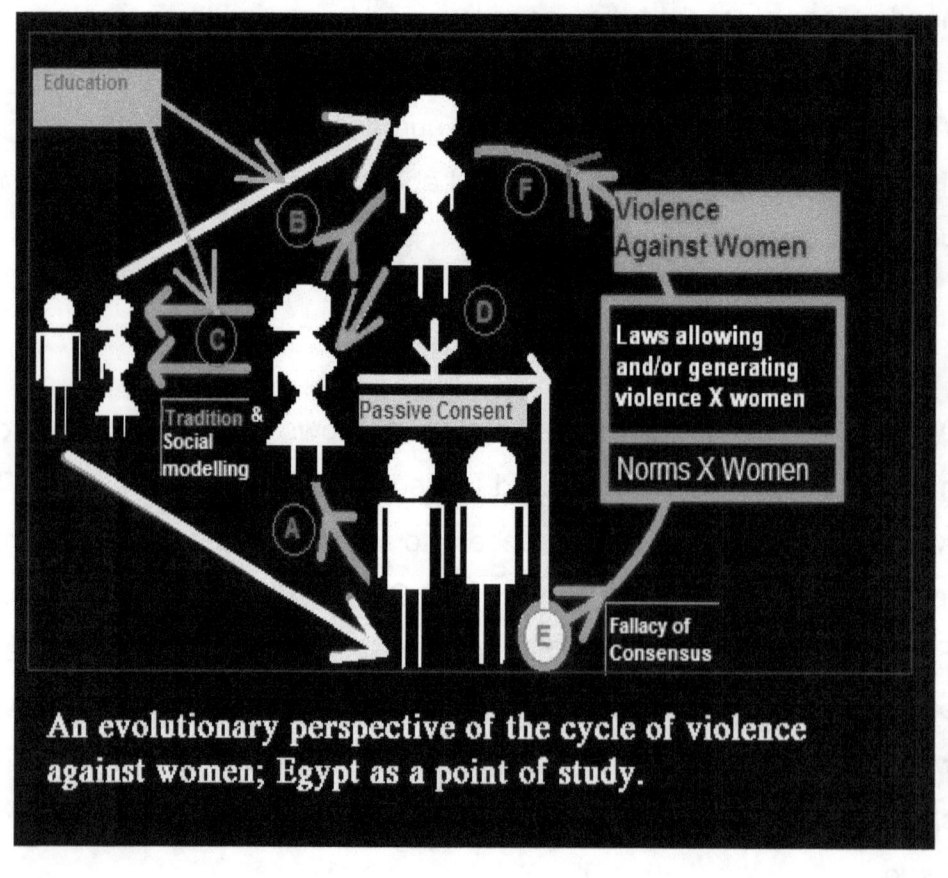

An evolutionary perspective of the cycle of violence against women; Egypt as a point of study.

Illustration 7: The Cycle of violence against women

I believe that the prevalence of violence against women is a major factor underlying much of poverty and hunger in the world, firstly, in the sense that abused women suffering Stockholm syndrome, may tend to raise aggressive male offspring for the "possession" of security and protection. I've seen many times a mother scolding a tearful son or getting bitterly sarcastic to deter such a "girlish" and quite often women get aggressive towards each other for such possession of a protective male; as in the case of Female Gentile Mutilation practised by women against women to patronize men. I've also seen a shocking image circulating in the social media of a newly born male wrapped up in a belt of bullets, so as to foreshadow bravery and dedication to aggression. The baby's mother must have been involved in the ceremony either by the active preparation or by subordination. Secondly, that subjection of women is losing a huge portion of creativity in businesses and political life, and finally, that subjection of women is of a greater political benefit to authoritarian societies and cultures than any criticism or lose incurred specially that women's rights rarely get advocated in relation to democracy due to its inconspicuous nature which is likely to perpetuate the cycle of violence and waste chances for productive cooperation among people beyond gender biases..

Taking Egypt as an example, being the worst Arab country for women to live' (report of the Independent, 2013), we may be able to see how the loop works:

Abusive norms (such as that woman are deficient "in mind" and thus should not be allowed to lead), abusive laws (such as ones allowing domestic violence towards women and honour killing), abusive economy and media reinforcement of submissive patterns, all interact to perpetuate the cycle of violence.

In 2011, a constitution that allows sexual-slavery and child marriage passed in spite of the resistance. Calls to re-legalize the FGM were brought to the parliament and started reverberating across the country through the daily discourse, social media and the media. We protested, then the constitution was replaced later by the new regime and public discussion on the FGM died, yet the same old laws remain abusive and discriminatory.

In addition to this, the unreported abuse incidents against women and female children make the statistics deceptive and an offensive community looking more protective of women than a liberated one.

The observation of the on-going violence against women in Egypt as a point of study proposes the long-term development of group Abuse

Syndrome at the point (D) of the violence loop. That is to say that many oppressed women may be resistant, at this point, to equality-education, as it threatens their fake security (Friere, 1993).

**Leaving Egypt** to take a *general look* on systems oppressive of women, Darwin has foreshadowed such phenomenon in his book The *Descent of Man*, by recording his own observation of the some primeval nations: at that time:

As (savage) do not regard the opinion of their women, wives are commonly treated like slaves. Most savages are utterly indifferent to the sufferings of strangers, or even delight in witnessing them. .Some savages take a horrid pleasure in cruelty to Animals,*(2) and humanity is an unknown virtue. Nevertheless, besides the family affections, kindness is common, especially during sickness, between the members of the same tribe, and is sometimes extended beyond these limits... Many instances could be given of the noble fidelity of savages towards each other, but not to strangers... This, again, is one of the virtues which becomes so deeply rooted in the mind, that it is sometimes practised by savages, even at a high cost, towards strangers; but to lie to your enemy has rarely been thought a sin, as the history of modern diplomacy too plainly shews. As soon as a tribe has a recognized leader, disobedience becomes a crime, and even abject submission is looked at as a sacred virtue... We have now seen that actions are regarded by savages, and were probably so regarded by primeval man, as good or bad, solely as they obviously affect the welfare of the tribe,- not that of the species, nor that of an individual member of the tribe

Such duality of within-group aggressiveness and compassion, while being blatantly cruel and deceptive with "strangers" and the subjection of women, were "natural" at a certain level of humanity evolution, however, with global citizenship calls and immigration trends and with our ethical evolution, such behavior can no longer be considered as natural today, especially if adopted as a code of conduct.

Such systematic selfish behavior at this stage of our ethical evolution would, according to Darwin's logic, occur either due to the lack of the "simplest reason" or as a very intelligent and well-informed strategic planning towards seizure of power in exploitation of today's global tendency to protect ALL that is labeled "culture".

As man advances in civilization, and small tribes are united into larger communities, the simplest reason would tell each individual that he ought to extend his social instincts and sympathies to all the members of the same nation, though personally unknown to him. This point being once reached, there is only an artificial barrier to prevent his sympathies extending to the men of all nations and races (Darwin, 1871)

Education for sustainable development in this particular respect has no magic wand; it has been assimilated to the' birth of a child; and a painful one' (Friere, 1993) as it intervenes to repair a previous education-of-oppression (i.e. oppressive conditioning), that becomes even more influential through the complexity of instruction (formal, non-formal and social modelling) and the sophistication of communication.

Unfortunately, there is no short-cut for freedom; education- for-equality needs to patiently empower, rather than to instruct, the oppressed towards freedom. But also significant political and legal interventions to address violence against women in terms of democracy are required.

Empowering and contributing to building such liberation awareness, require weaving the fundamental concepts of freedom, equality, solidarity, perceptive tolerance, respect for nature, and shared responsibility into schools curricula and extracurricular activities in addition to all lifelong learning programs with a special attention to the education of immigrants as they strive to adapt to new societies. It also requires activism of continuous communication of such values in our daily life and social interactions and above all, it requires a human unity of self-reference.

## 8.3- POVERTY & ENERGY: RESOURCES OR SYNERGY CRISIS

As a consequence, the idea of *autopoiesis*, when applied as an instrument of social analysis, confirms the conclusion already established by other means of investigation — that our societies are self-mutilating, pathological systems (Maturana and Valera, 1928).

A precise evaluation of the situation is currently unachievable due to the missing wholeness of our self-organization, yet, it is easy to draw on some simple facts to depict a conceptual sketch of a required initial synergy required -in turn, for population self-organization:

While there is an estimated number of 700 million of hungry people around the world (WFP, 2014), roughly one third of the food produced in the world for human consumption every year gets lost or wasted according to the FAO. Moreover, according to Women in Agriculture (2011)"If women farmers had the same access to resources as men, the number of hungry in the world could be reduced by up to 150 million". This is in addition to that much of today's hunger seems to be caused by extraordinary conditions such as wars, civil wars, destruction of infrastructure then hunger traps (according to the

Challenges of Global Poverty course provided by MIT through EdX, 2013: 2014).

We are complaining of the depletion of fossil fuel and/or global warming, for example while the little bit of sunlight falling on earth for just one hour meets world's energy demands for an entire year! (Maclamb, 2010).

How much of today's hunger problem can be solved by self-organization when the world produces enough food to feed each individual on the globe, yet one person in eight doesn't have enough food to maintain one's life?

Energy and matter are naturally, inter-reversible (Umpleby, 2007). The idea of "eating" to live is an idea of transforming matter into energy and then energy into matter plus an ability to exert an effort. This is an amazing ability of all living systems. Human beings due to the unique intellectual ability (inclusively speaking) are able to manipulate all elements of nature through Cybernetics skills, to mobilize energy and matter into their self-organization with a potentially infinite capacity(illustration 2 in relation to illustration 6). That is that we might not be having an energy nor poverty crisis; but globe-population-self-organization creation crisis.

This, might suggest that, the only "within-group competition" we might really have to undergo, is the attempt to assist nature in eliminating the malignant (not the physically-weak)individuals and/or groups(a point specifically visited in section 7.1, lines 965-990; section 7.2, this section, and extended in sections 8.4 and 9) through justice system, international cooperation and perceptive tolerance; where education(in all its forms) serves as the intrinsic communication system of our globe-population-self-organization.

There are then many alternative resources of energy and food, and cure to almost all the harmful substances human beings produce into the environment, if we were more aware of the manipulation process. One of such amazing alternatives is "mushroom"(Miller K., 2013): the "forgotten kingdom" (Fungus Conservation Forum, 2008; Miller 2013. Energy and matter exist abundantly in nature but once a form of organismic life has been wasted, it cannot be brought back.

Education for sustainable development- does not "teach" solutions but communicates values and skills that will produce solutions (Soliman et al, 2013) out of the educational process (illustration-4) (Soliman et al, 2013).

Education for sustainable development, in this respect, does not concern itself with the number of educated people around the world only but rather the quality of education provided.

What if education IS the problem?

## 8.4 - NEUTRALIZING THE "WRONG"PACKETS OF INFORMATION

Human beings have this capacity of observing not only things but "no-thingness" of things.(Kenny, 2009). This makes us able to acknowledge concepts such as the anti-matter as the absence of matter (precisely a matter with a negative charge; NASA, 2008), death as the absence of life, darkness as the absence of light and any "missing "component/person that might have otherwise contributed to the success of any process or situation. This detection of "nothingness" or missing components is essential for strategic planning and is a ubiquitous part of Cybernetics.

By observation, we know that there are some packets of information driving individuals and groups in our communities that are missing the love component (altruism traits).

Such packets of information define selfish behaviour and may *occur* naturally in other species and kingdoms of life, while for the species of *Homo sapiens*; the nature's Cyberneticists, they get consciously *adopted.* Moreover; and more seriously, they may also get developed into "moral code" for a group of selfish individuals cooperating temporarily in achieving dominance over resources with no intention to distribute benefit or remain in solidarity after mission is accomplished.

The prevalence of moral systems driven by moral codes missing the love value, or imposing a temporal within-group sort of altruism, assimilates the prevalence of "cheats" in the case of *Pseudomonas aeruginoshe* and cancerous cells (Piliavin & Charng, 1999) in the human body. This may include dangers from the artificial intelligence (BBC news, 2013), inconsiderate capitalism and terrorism.

Groups of genes form genomes, which form cells, which form multicellular organisms, which may form social groups and so on.

Multicellular organisms comprise populations of cells, which cooperate to enhance the fitness of the whole. However, as with all evolutionary transitions, there is the ever-present temptation to cheat so that fitness at the lower level is enhanced at the expense of the higher level. 'Cheaters' abort their altruistic behaviour at the higher level in favour of selfish behaviour and a fitness benefit at the lower level. This process is the basic pathology of cancer, albeit that the benefit to cheaters may be transient given that the lower-level unit's survival is usually dependent upon the survival of the higher-level unit (although there are at least two exceptions to this norm, see Murgia et al.21 and Siddle et al.22).

Moral systems that are indifferent to the form of life in hand, providing non-pro-life justice standards for a life organization(s), might be a basic barrier in the way of our unity and self-organization. Such moral systems adopting moral codes that punish brutally and recommend ending the life of individuals for non-anti-social reasons such as embracing different beliefs or sexuality, or constantly provoking the "us" and "them" among classified groups are - dogmatically anti-social and thus, for humanity, "self-mutilating" (Mariotti, OIKOS) and pathogen.

Unlike other species and kingdoms of life, maladaptive individuals are neither the weak or the sick nor the poor, because our intellectual evolution has made up for almost all our physical weaknesses (illustration-12); maladaptive individuals are as Darwin describes; ones lacking compassion for humanity and the power of reason as well as self-commandment; those described by Darwin (1999) as of "low-morality".

I propose that the "survival of the fittest" in our intellectual realm translates into the (collective) survival of the *Wise. Actually, the very name that lumps us up with other animalls is the one the sies us in distinction: Homo  sapien or  as in Greek the wise man.*

Education for sustainable development whether formal or non-formal is potentially capable of neutralizing pathological systems, by communicating values of cooperation, humanity but also of justice. That is to reinforce *mutual* tolerance patterns correspondent to the need for both love and justice and protective of the kind/cooperative individual, thus of all humanity.

## 9- *THE* OBSERVER

It's impossible to go through ecological ethics debate without facing a failure to locate the position of the human kind to nature. In an attempt to reach a balanced position that would work for environmental ethics, Peterson (1999) suggested a sort of "productive tension between realism and constructivism". For constructivists, nature is appreciated as a cultural product. It is either structured exclusively through linguistic interpretation of the correspondent natural objects, or *physically* structured by and through human cultures. In terms of cybernetics, that if energy and matter are inter-reversible and information is a "reduction of uncertainty" (Bateson, 1972) that maybe this world does not really exist. Naturalists, on the other hand, regard human beings as one species among millions of other living organisms and natural objects contributing to the structure of life on earth. Peterson did not challenge any of such "appreciations" in principle but rather in praxis: Adopting any of such views in their extremes - as she notes - leaves no space for environmental ethics. One of them, respectively, is reductional of species and landscapes while the other might deem all ethical choices as morally "good" in a "natural fallacy". The solution according to Peterson would be a "constrained" realism.

Such "constrained realism" principle as attractive as it might seem, is incapable of justification through a mono-speciality perspective. It leaves the reader wondering if we should structure ethics for nature or nature for ethics! Reduction is inevitable, for data transmission and communication (Umpleby, 2007), but it is unrealistic to assume that by reduction humans become the sole author of nature nor a mere decoder.

As far as ecological ethics are concerned, natural forces, processes and entities remain as real as they can set constraints and even decompose the observer. and as real as they can manipulated and simulated through observation within a spatiotemporal frame of mutual action. The battle then is real enough between the natural forces and the vulnerable observer and thus what might work best in decision making is a realistic constructivism admitting a level of certainty (science) and a level of uncertainty represented by the coefficient [h] in any system.

The forces of natural laws such as gravity; elements of space, time, pressure and temperature; chemical reactions producing energy; matter-and-antimatter-annihilation producing energy, the random mutations, information ebb and flow manifested in system formations are the world as we perceive it, whether regarded - in their circularity - as a Providence or just what they appear to be: The "aggregation of natural laws and processes". But as any *"Gestalt – perceivable whole"*, such aggregation must be greater than its fragments introducing us to a hypothetical Observer as circularity requires (Umpleby, 2007) which is possibly why Darwin, in spite of stressing that he did not think of nature as a power or a deity, found a difficulty not to personify "her". Whether this observer is an intelligent power is something beyond science to affirm or deny, but we can suppose that by the compliance with such aggregation of enforced laws and by respecting other forms of life that we could reach a level of harmony with such suprasystem(s) and thus reach the ultimate wisdom. It might be interesting in this respect to note that the name which lumps humans up with the rest of animal species is the very name that sets us apart from all other beings: The *Homo sapiens* or the wise man. In all cases, this remains as a mystical area in cybernetics and there is nothing that can be done about that.

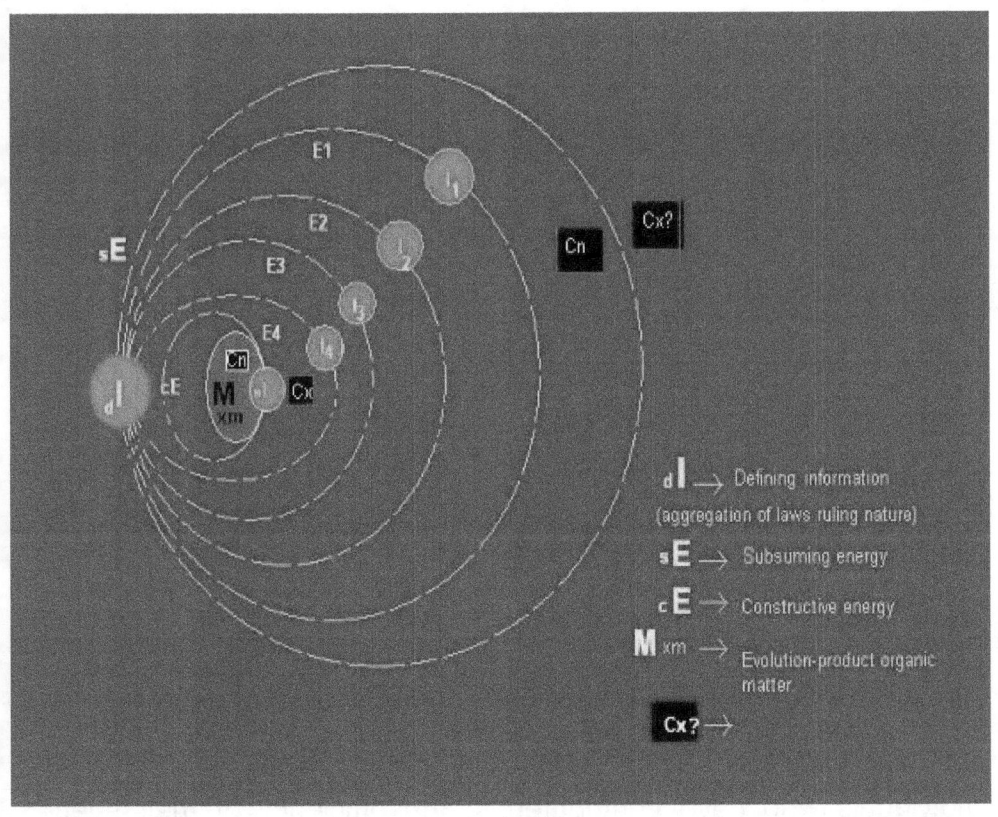

**Illustration 8: A perspective of a prospective harmony with nature as The HhIECM**

Darwin proposed that species (as systems) has a capacity each. That organisms *cannot* all survive *because* then they would run out of space and resources, giving the example of one pair of elephants, as the "slowest breeders on earth" who would produce 19 million elephants in about 750 years if the natural selection didn't destroy good deal of their offspring.

Natural selection works through preservation of the adaptive qualities, once they occur, and the destruction of the maladapted individuals so as - as inferred by Darwin's logic - to reserve space and resources for the well adapted individuals and species. This is a species- capacity unarticulated proposal.

This means that reproduction itself cannot pass, by nature, as a continuation of life or an indicator for species success.

A current population of a species would be the representation of their progenitors and would be represented in their progeny only if the given species maintains existence through evolution. Maladapted populations just perish (go extinct), decomposing back into earth to feed and sustain the surviving species. And in this sense the individuals who have contributed to the efficiency of their species, remain in the *phylogeny* of all the coming generations as long as the latter maintains existence. Species cannot realize this however; "They are just what they are" (Glasersfeld, 2002)! In the case of human cognitive evolution; those individuals are the ones who contribute to the welfare of humanity by promoting ethics, wisdom, synergy and any sort of constructive knowledge. And in this sense, any word uttered for the welfare of humanity when preserved in artefacts, is part of our evolutionary development and product,

This sense is expressed in the proposal that nature rules by "killing" (Glasersfeld, 1980) the maladaptive individuals and species, while is said to be working for the "good of every creature" at the same time. "Every creature" here is obviously an inclusive term referring to "each" intergenerational population.

Education in this respect would be the most effective way known to produce a "progress in corporeal and mental endowments" (Darwin, 1871) as an attempt to extend existence beyond biological life-time span of individuals parallel and complementary (or alternative in individual cases) to biological reproduction processes.

A tribe including many members who, from possessing in a high degree the spirit of patriotism, fidelity, obedience, courage, and sympathy, were always ready to aid one another, and to sacrifice themselves for the common good, would be victorious over most other tribes; and this would be natural selection (Darwin, 1999).

This is how Darwin expressed the "Survival of the Fittest" paradigm but then in compliance with the fundamental objectivity principle, scientists later ignored such holistic perspective and sought system controls in their fragments till a holistic Darwinism (Corning, 1997) has been finally acknowledged.

Our tribe in this respect would be humanity as Darwin confirms later in the same discourse and as he proposes in the Origins of Species; that nature does not seek perfection through the ultimate preservation of species but through allowing them to advance in organization.

This improvement inevitably leads to the gradual advancement of the organisation of the greater number of living beings throughout the world (Darwin, 1999)

Corning(1997) emphasises thus the need for a "purposeful innovation" as a source of creativity in interpreting the evolution process rather than mutations and recombination mechanisms.

## 10- THE *Homo Cyberneticist*

I propose that either *Homo sapiens* get listed at the top of the IUCN red list of endangered species for the deficiency of "social instincts" and the potential resulting chaos; or that it is may be high time human beings get set as a separate *sociophysiobiological* kingdom based on the accumulation of evolutionary cybernetic codes preserved in scripts and models of self-regulatory system allowing the mobilization of energy and matter to modify or supplement our biological existence and enhance our adaptability by expanding our organizational capacity.

That is in addition to the discovery of Lateral frontal pole prefrontal cortex area, in the ventrolateral frontal cortex region of the brain, which has been observed as uniquely human and which might have produced the evident gap of civilization between human beings and all other beings on earth.

**The Cybernetic Phylogeny**:

hICEM = h * {I.a + I.a~b + I.a~b~c + I.a~b~c~d + I.(cognitive code [vehicle]) + I.(cognitive code [aeroplane]) + I.(cognitive code [self-defence-weapon])+ I.(cognitive code [submarine]) + I.(cognitive code [TV]) + I.(cognitive code [agriculture]) + I.capacity+I.z} + {[Cn]+[Cx]+[C±]} + {E1+E2+E3+E4+E5+E6+ E7+E8+E9+E±}±{M1XM2XM3XM4~(car) X M5(aeroplane) X M6~(self-defence-weapon) X M7(submarine) X M8~(TV) X m9~(Farm) X [M.capacity] X M.z} *

-------------------------------------------------

Where a, b, c are genetic modification and 1, 2, 3 are biological modification involving an "attached" physical and social phenotype correspondent to attached cognitive codes (scientific-knowledge heritage).

# 11- CONCLUSION

This is a pragmatic framework relating education to environmental ethics through a cybernetic perspective. The study proposes that the third order cybernetic (Kenny, 2009) is cybernetics of providence; of observing one's own organisation and address the gap of awareness among observers which is required for a harmonized decision-making processes and thus system synergy. This is a process of self-creation as it combines system dynamics, system management and system creation approaches. It is different from totalitarianism or social engineering as it depends on the existence of a mechanism for mutual persuasiveness (democracy) and "mutual coercion" (Hardin, 1968) to ensure the system viability.

Education, in this respect, may be defined as the intrinsic communication system of our life organization, once such organization has been established through education. This sets quality education and sustainable development in a circular causal relationship. It also shows that quality education and integrated management are fundamental mechanisms for achieving sustainable development..

This research recommends communicating a association of compassion/love/tolerance and social justice through all our systems - as the core values underlying the discriminative nurture (properly interpreted) processes dominating nature – with a vision to unite humans beyond borders and biases.

This perspective does not rule out the possibility of the existence of a Providence but does not confirm it either. There is an amount of uncertainty which has to persist in regards of this issue. The study warns, however, against the attempts to promote ultimate tolerance and acceptance of all cultures while the world governance systems are still based on bitter division, politically-arranged inequality and culturally sanctioned violence. It also recommends addressing gender equality in terms of democracy rather than feminism.

This study highlights the *spaciotemporal* dimension of ecosystems in relation to an observer. Natural processes and entities are as real as they are able to suffer or cause suffering, set constraints in the way of, or even to decompose the observer, at a certain point in time and space. They are also as real as they would be manipulated towards equilibrium or chaos by an observer.

Nature for us is about the here-and-now stamped by the past and reaching out for a future.

In the Conservation of *Homo sapien* I finally propose that either *Homo sapiens* get listed at the top of the IUCN red list of endangered species for the lack of social instincts and the prospective chaos or that it is maybe high time human beings are set as a separate *sociophysiobiological* kingdom based on our cybernetic phylogeny encompassing all elements of nature and other species through the capacity of observation. Realizing such capacity might urge humans to unite beyond ethnicities and gender biases to be well represented in the next generations; Our children and grandchildren.

Reference List:

Bandura (1971) *Social Learning Theory.*Standford University, General Learning Corporation.

BBC News (2013) *How are humans are going to become extinct*? Available at [http://bbc.co.uk/news/business-22002530 ]

BBC News(2013) *What is Stockholm syndrome*?, BBC News Magazine available at [http://www.bbc.co.uk/news/magazine-22447726]

Beer, S. (1972) *Brain of the firm/by Srafford Beer*: a development in management cybernetics/by Stafford Beer. The Professional Librarywith Penguin Press.

Bandura (1971) Social Learning Theory Standford University, General Learning Corporation.

Corning P. (1997) A Holistic Darwinism "Synergistic Selection"and the Evolutionary, Process,.Institute for the Study of Complex Systems, JAI Press

Darwin, C. (1871) *the Descent of Man*, and selection in relation to sex, New York: D. Appleton.

Darwin, C. (1999) *On the Origin of Species by means of Natural*

*Selection*, 6th Edition, available a[http://www.classicly.com/charles-darwin/the-origin-of-species-by-means-of-natural-selection-6th-ed/download/pdf]

Elliot, R. (1982) *Faking Nature* Inquiry 25 : 81-93; reprinted in Donald VanDeer and Christine Pierce, eds, people, penguins, and plastic Trees: Basic Issues in Environmental Ethics( Belmont, Calif: Wadsworth, 1986), pp. 142-50

Encyclopedia Britannica . Phylogeny. Available at[http://www.britannica.com/EBchecked/topic/458573/phylogeny]

Encylopedia Britannica. Homo sapiens. Available at [http://www.britannica.com/EBchecked/topic/1350865/Homo-sapiens]

Encyclopedia Britannica. *Endangered species*. Available at[http://www.britannica.com/EBchecked/topic/186738/endangered-species

Foerster H.V. (1979) *Cybernetics of Cybernetics*, University of Illinios, Urbana.

Gallopin G. (2003) *A systems approach to sustainability and sustainable development Sustainable*. Development and Human Settlements Division ECLAC/ Government of the Netherlands Project NET/00/063 "Sustainability Assessment in Latin America and the Caribbean", 64 Santiago, Chile, March.

Gray P. ( 2011) *The Evolutionary Biology of Education*: How Our Hunter –Gatherer Educative Extincts could form the Basis For

Education Today, Springer Science+Business Media, LLC 2011.

Beer, S. (1972) Brain of the firm/by Srafford Beer: a development in management cybernetics/by Stafford Beer. The Professional Librarywith Penguin Press.

Brady, E. (2007) *Aesthetic Regard for Nature in Environmental and Land Art, Ethics, Place &Environment*: A Journal of Philosophy &Geography, 10:3, 287-300

Ceccarelli, L. (2001) *Uniting Biology and the Social Sciences:*A Rhetorical Comparison *of E. O. Wilson's Consilience and Theodosius Dobzhansky's Mankind Evolving. Poroi 1, Iss. 1 (2001).Available at: [http://dx.doi.org/10.13008/2151-2957.1005]*

*Carlson, A (2010) Contemporary Environmental Aesthetics and the Requirements of Environmentalism*. Environmental Values 19 289–314. © 2010 The White Horse Press
doi: 10.3197/096327110X519844

Condillac, E.B. (2001) *Essay on the Origin of Human Knowledge*, translated and edited by Hans Aarsleff , Princeton University, Cambridge University Press.

Corning P. (1997) *A Holistic Darwinism "Synergistic Selection"and the Evolutionary Process.*Institute for the Study of Complex Systems, JAI Press

Curzon, L.B. (1990) *Teaching in Further Education*, fourth edition

Darwin, C. (1871) *the Descent of Man, and selection in relation to sex*, New York: D. Appleton.

Darwin, C. (1999) *On the Origin of Species by means of Natural Selection,* 6th Edition, available at [http://www.classicly.com/charles-darwin/the-origin-of-species-by-means-of-natural-selection-6th-ed/download/pdf]

Department of Environment Food and Rural Affair (2013) *Sustainable Development Indicators,* Sustainable Development Statistics, The National Archives, Kew, London, available at [https://www.gov.uk/government/organisations/department-for-environment-food-rural-affairs/series/sustainable-development-indicatorsv].

Diehm, C. (2013) *Wolves, Winsconsin, and Aldo Leopold. Minding Nature.* Reviews &reflection.Vol. 6, no.2 Journal of Philosophy

Doebeli M. and Hauert C. (2005) *Models of cooperation based on the Prisoner's Dilemma and the Snowdrift game,* Ecology Letters, 8:748-766

Elliot, R. (1982) *Faking Nature Inquiry*25 : 81-93; reprinted in Donald VanDeer and Christine Pierce, eds, people, penguuns, and plastic Trees: Basic Issues in Environmental Ethics( Belmont, Calif.: Wadsworth, 1986), pp. 142-50

Encyclopedia Britannica . Phylogeny. Available at[http://www.britannica.com/EBchecked/topic/458573/phylogeny]

Encylopedia Britannica. *Homo sapiens.* Available at [http://www.britannica.com/EBchecked/topic/1350865/Homo-sapiens]

Encyclopedia Britannica. *Endangered species*. Available at [http://www.britannica.com/EBchecked/topic/186738/endangered-species

Espejo R. (2003) *The Viable System Model*: A Briefing about Organizational Structure, SYNCHO Limited, Website: www.syncho.com.

Excellent Gateway Website [available at: http://www.excellencegateway.org.uk/node/12016], [accessed: 23/11/2013] Defining teacher roles and responsibilities in the further education sector in England, 2008.

Featherston, J. *Cooperation and conflict in cancer: An evolutionary perspective*. S Afr J Sci. 2012; 108(9/10), Art. #1002, 7 pages, v108i9/10.1002 available at[ http://dx.doi.org/10.4102/sajs]

Foerster H. V. (1979) On *Cybernetics of Cybernetics*, University of Illinois, Urbana

Friere, P. (1993) *The Pedagogy of the Oppressed*, New York: Continuum Book.

Fungus Conservation Forum (2008) *The Forgotten Kingdom: A Strategy for the Conservation of the UK's Fungi*: 2008-2015, published by Plantlie International on behalf of the Fungus Conservation Forum.

Gallopin, G. (2003) *A systems approach to sustainability and* sustainable development Sustainable.Development and Human Settlements Division ECLAC/ Government of the Netherlands Project NET/00/063 "Sustainability Assessment in Latin America and the Caribbean", 64 Santiago, Chile, March.

Gray P. ( 2011) *The Evolutionary Biology of Education : How Our Hunter – Gatherer Educative Instincts could form the Basis For Education Today*, Springer Science+Business Media, LLC 2011.

Gardiner, S. (2006) *A Perfect Moral Storm*: Climate Change, Intergenerational Ethics and the Problem of Moral Corruption.

Glanville, R. (2004) *The purpose of second-order cybernetics*. Kybernetes. Emerald Group Publishing Limited Vol. 33 No. 9/10. pp. 1379-13860368-492X. DOI 10.1108/03684920410556016

Glasersfeld, E.V.(1980) *"Viability and the concept of selection."*, American Psychologist (vol.35, 1980, 970–974)

Glasersfeld E.V. (1996) *Cybernetics and the Art of Living*, 13th European Meeting on Cybernetics and Systems Research, Vienna,April 922-

Glasersfeld E.V. (2002) *Cybernetics and the Theory of Knowledge*, UNESCO Encyclopedia

Gott, J.M. (2003) *Expanding genome capacity via RNA editing.* Comptes Rendus Biol., *326*, 901-908.

Gravells, A. (2007). *Preparing to Teach in the Lifelong Learning Sector,* Third Edition Exeter: Learning Matters.

Hardin, G. (1968) *The tragedy of the commons.*

Science162:1243–48. [CW] cited in

Harp &Raw (1977) *Carl Roger and Humanistic Education*, Chapter 5 in Patterson, C.H. Foundation for the Theory of Instruction and Educational Psychology

Higgins, L (2006) *Some Notes on Assessment Theory*, TLC Project Partner. Southampson Solent University.

Holmes G and Abington-Cooper M., Pedagogy vs. Andragogy: A False Dichotomy?, The Journal ofTechnology Studies, Volume 26, Number 2

Jones and Nemeth (2005) *Cognitive Artifacts in Complex Work*, Redesign Research, USA, The University of Chicago, USA

International Council for Science, 2005. *Harnessing Science, Technology, and Innovation for Sustainable Development*. A report from the ICSU-ISTS-TWAS Consortium ad hoc Advisory Group.2005

Johnson &Raven (2002), *Biology*, McGraw-Hill Higher Education..

Katz, E. (1997) *The big Lie: Human Restoration of Nature.*

Katz, E. (2012) *Further Adventures in the case against restoration*Environmental Ethics.

Learning-Theories.Com Website: *Maslow's Hierarchy of Needs*, available at [http://www.learning-theories.com/maslows-hierarchy-of-needs.html], accessed on: 23/01/2014

Lederberg, E, and J. Lederburg (1952) Evolution 101, available at:[http://evolution.berkeley.edu/evosite/evo101/IIIC1aRandom.sh

tml]

Lifelong Learning UK (2006) *New overarching professional standards for teachers*, tutors and trainers in the *lifelong sector, www.lifelonglearninguk.org*

*Live Science (2014) Newly Discovered Brain Region Helps Make Humans Unique*, By Tia Ghose, Staff writer, available at [http://www.livescience.com/42897-unique-human-brain-region-found.html], accessed on 2/4/2014

Lunenburg F. C. (2011) *Curriculum Development: Deductive Models*, Sam Houston State University, SCHOOLING VOLUME 2, NUMBER 1, 2011

Light, A. (2011?) *Ecological restoration and the culture of nature*: a pragmatic perspective. Eirie Baffalo Edu

Maclamb (2010) *The Secret World of Energy, Ecology Global Network*available at [http://www.ecology.com/2010/09/15/secret-world-energy/], accessed on: 1/3/2014

Mariotti Humberto (OIKOS), *Autpoiesis, culture, and society*available at[http://www.oikos.org/mariotti.htm]

Martincorena I, et al (2012), *Evidence of non-random mutation rates suggests an evolutionary risk management strategy*, Nature publishing group

Maturana and Valera (1928) *Autopoieses and Cognition*: The realization of the Living,Boston studies in the philosophy of science;       v.42),       available       at

[http://topologicalmedialab.net/xinwei/classes/readings/Maturana/ autopoesis_and_cognition.pdf]

McLeod, S.A. (2007). *Maslow's Hierarchy of Needs*. Available at:[http://www.simplypsychology.org/maslow.html]

Merriam Webster Online Dictionary (2014) *Cybernetics*, Merriam Webster, Inc. Available at [http://www.merriam-webster.com/dictionary/cybernetics ], accessed on; 5/7/2013

Mindell, D (2000) *Cybernetics: Knowledge domains*in Engineering systems.

National Aeronautics and Space Administration, Newton's Laws of Motion, available at [http://wwww.grc.nasa..gov/WWW/k-12/airplane/newton.html], accessed on 12/3/2014.

Naess, A. (1973) The shallow and the deep, long-range ecology movement. A summary. Inquiry: An Interdisciplinary Journal of Philosophy, 16:1-4, 95-100 available at: [http://dx.doi.org/10.1080/00201747308601682]

Nowak et al (2010) The evolution of Eusociality Macmillan Publishers Limited.

O'Neil d (1998-2013)Early Theories of Evolution: Darwin and Natural Selection, Anthro-palomar available at [http://anthro.palomar.edu/evolve/evolve_2.htm]

O'Neil d (1998-2013)Early Theories of Evolution: Darwin and Natural Selection, Anthro-palomar available at [http://anthro.palomar.edu/evolve/evolve_2.htm]

Ophardet, C.E. (2003) Virtual Chemobook; *Types of RNA*, Elmhurst College available at: [http://www.elmhurst.edu/~chm/vchembook/580DNA.html].

Palmer, C.(2003) Placing Animals in Urban Environment. Journal of Social Philosophy, Vol. 34 No.1. Blackwell Publishing, Inc.

Park, J.H.(2007) *Persistent Misunderstandings of Inclusive Fitness and Kin Selection: Their Ubiquitous Appearance in Social Psychology Textbooks.* Evolutionary Psychology – ISSN 1474-7049 – Volume 5(4). 2007. -861-available at[http://www.epjournal.net/wp-content/uploads/EP05860873.pdf]

Parra-Luna F, (2009), *Systems Science and Cybernetics: The Long Road to World Sociosystemicity-* Francisco Parra-Luna ©Encyclopedia of Life Support Systems (EOLSS)

Patterson, C. H. (1997) Foundation for the Theory of Instruction and Educational Psychology, Harp &Row.

Peel D. (2005*) The significance of behavioural learning theory to the development of effective coaching practice*, Hartnell Training Ltd. UK, International Journal of Evidence Based Coaching and Mentoring Vol. 3, No. 1, Spring 2005.

Peterson, A. (1999) *Environmental Ethics and the Social Construction of Nature.* Environmental Ethics. Vol. 21 subvert

Piliavin &Charng, *A review of Recent Theory and Research*, Annual Review of Sociology, Vol.16 (1990), pp.27-65, Annual Reviews, available at[http://www3.nd.edu/~wcarbona/piliavin-altruism-ARS.pdf] accessed at 23/2/2014

Preparing to Teach in the Lifelong Learning Sector (*LLUK*

*Professional Standards*), Learning Matters, Third Edition.

Preston, C. J. (2012) Beyond the End of Nature: SRM and Two Tales of Artificity for the Anthropocene. Ethics, Policy and Environment. Vol. 15, No. 2

Pross A (2011) *Toward a general theory of evolution: Extending Darwinian theory to inanimate matter,* Pross Journal of Systems Chemistry, vailable at [ahttp://www.jsystchem.com/content/2/1/1]

Rachlin H (2002).*Altruism and selfishness* Behavioral and Brain Sciences 25, 239–296

Reece, I. &Walker, S. (2003, 2007) *Teaching, training and learning,* (5th-6th editions). and Sunderland. Business Education Publishers Ltd.

Ridly, M (1995) *Animal Behaviour,* Blackwell Scientific publications

Robock, A. (2008) *20 Reasons why geoengineering may be a bad idea:*Bulletin of the Atomic Scientists MAY/JUNE (Vol. 64, No. 2, p. 14-18, 59 DOI: 10.2968/064002006

Ruggiero et al. (1994) *Population Viability Analysis, Corel Corporation Limited,* Available at[ http://warnercnr.colostate.edu/~gwhite/pva/], accessed on 4/2/2014.

Schroeder D.(2005) *Evolutionary Ethics,* Lancaster University, United Kingdom, (Internet Encyclopedia of Philosophy).

Schwaninger M. (2004) *System Dynamics and Cybernetics: A Necessary Synergy*, International System Dynamics Conference, Oxford, July 2004.

Smith, M.K. (1996, 2000) *Curriculum theory and practice*, the encyclopedia of informal education, available at [www.infed.org/biblio/b-curric.htm.

Singer, P. (1989) All Animals Are Equal.In Tom Regan &Peter Singer(eds.) Animal Rights and Human Obligations. New Jersey: Prentice-Hall. pp. 148-162

Soff, M. (2013) *Gestalt Theory in the Field of Educational Psychology*: An Example, Gestalt Theory, Vol. 35, No. 1

Soliman, G.S. *SSCC Project*, 2013, MIT Forum Enterprise; Semi-finalist, Team 46.

Stafford Beer(Uploaded on 19 Jun 2009) *a Viable System Model ; Watt's Engine*, available at: [http://www.youtube.com/watch?v=q3yNJPkdtYo].

Stanford Encyclopaedia of Philosophy (2003) *Biological Altruism*, Available at [http://plato.stanford.edu/entries/altruism-biological/ ], accessed on: 11/24/13

Thames Water Website, 1800-1900- The Great Stink [avilable at ] accessed on 23/11/2014

The Independent (2013) *Revealed: Egypt is the worst Arab country for women*, available at [http://www.independent.co.uk/news/world/africa/revealed-egypt-is-the-worst-arab-country-for-women-8933608.html ] accessed on 12/11/2013

The IUCN Red List of Threatened SpeciesTM (2014). Available at [http://www.iucnredlist.org/details/136584/0]

68

The United Nations General Assembly (2000) *The Millennium Declaration*available                                                                    at [http://www.un.org/millennium/declaration/ares552e.htm]

Tummons J. (2010) *Becoming a Professional Tutor in the Lifelong Learning Secor* (Achieving QTLS Series) (p. 60). Learning Matters. Kindle Edition.

Turtle World Website Sea Turtle Reproduction, available at[http://www.seaturtle-world.com/sea-turtle-reproduction/]

Tylor, P.W. (2011) Respect for Nature: A Theory of Environmental Ethics.               Princetonwww.thameswater.co.ik/about-us/850-2611.htmUniversity Press

Umpleby, S.A. (2007) *Physical relationships among matter, energy and information*(Reprinted formCybernetics and Systems '04, 2004). Syst. Res. Behav. Sci. 2007, 24, 369-372.

UNEP (2009) *Global Food Losses and Food Waste*- FAO, 2011, the environmental crisis: The environment's role in averting future food crisis –

UNESCO, 1996 *Management of Social Transformations*(MOST) University Twinning Programmed (UNITWIN) Policy Paper - No. 3

Varese F.(2001)*Altruism and the Theory of Rational Choice: an Emperical Exploration*, in collaboration with Meir Yaish.

Vincent Kenny (2009) *"There is nothing like the real thing"*,Revisiting the Need for Third Order Cybernetics, Constructivist Foundations Volume4- Number 2.available at [http://www.univie.ac.at/constructivism/Journal], accessed on 24/2/2014

Walkin, L(2000) *Teaching and Learning in Further and Adult Education*, Stanely Thorne.

Wallace, S. (2005) *Teaching &Supporting Learning in Further Education*, Learning Matters.

West et al (2006) *Altruism*, Institute of Evolutionary Biology, School of Biological Sciences, University of Edinburgh Current Biology, Vol 16 No 13 R482.

West et al (2006) *Social evolution theory for microorganisms*, Nature Publishing Group, VOLUME 4

Wilson &Holldobler(2005) *Eusociality: Origin and consequences*, PNAS, vol. 102 no. 38 13369.

Wilson &Wilson (2007) *Rethinking the Theoretical Foundation of Socio-biology*, Quarterly review in Biology, press, Vol. 82, No. 4, December, available at [http://www.jstor.org/stable/10.1086/522809].

Wilson E. O.(1998) *Consilience: The Unity of Knowledge*, New York, Knopf, 1998, p. 13.

World Food Program Website, *Fighting Hunger Worldwide*, Hunger, Hunger Statistics; available at [http://www.wfp.org/hunger/stats] accessed on: 2/3/2014.

Weathreall, D. J. (2001) *Genotype-Phenotype Relationships*. Encycopedia of Life Sciences. NaturePublishing Group

West et al (2006) *Altruism*, Institute of Evolutionary Biology, School of Biological Sciences, University of Edinburgh Current Biology, Vol 16 No 13 R482